LIFE HACKS
FOR PARENTS

Summersdale Publishers Ltd
46 West Street
Chichester
West Sussex
PO19 1RP
UK

www.summersdale.com

Printed and bound in Croatia

ISBN: 978-1-78685-000-3

Substantial discounts on bulk quantities of Summersdale books are available to corporations, professional associations and other organisations. For details contact general enquiries: telephone: +44 (0) 1243 771107, fax: +44 (0) 1243 786300 or email: enquiries@summersdale.com.

LIFE HACKS
FOR PARENTS

Handy Hints to Make Life Easier

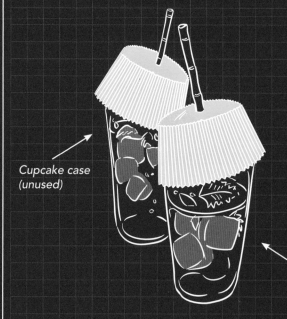

*Cupcake case
(unused)*

*Bug-free
beverage*

Dan Marshall

Over **130** amazing hacks inside!

summersdale

DISCLAIMER

Neither the author nor the publisher can be held responsible for any loss or claim arising out of the use, or misuse, of the suggestions made herein.

CONTENTS

INTRODUCTION

Welcome to *Life Hacks for Parents*, an indispensable guide to help you successfully traverse the ups and downs of being a parent. From home safety and first aid hacks to cleaning and food hacks, and everything in between, within this book you will discover over 130 ways to make your life that little bit easier, whether you have a newborn or a teen – or both.

Perhaps you need to remove some of your child's 'artwork' from the wall, or you want a natural way to soothe your baby when they're teething, or maybe you're just desperate for some fun activities for the holidays that won't render you bankrupt? It's all covered – and the best bit is, you'll most likely have the necessary ingredients at home already.

Life Hacks for Parents is here to take the strain out of parenting. So read on and learn how to be a hacking hero – your children will thank you for it.*

*If you're lucky.

EARLY DAYS HACKS

Newborn babies are magical and delightful, but nothing can prepare you for exploding nappies and being up all hours for feeds, not to mention the challenges of dealing with any of these things on the go! This chapter offers some handy hacks to help you through those fuzzy first few months.

ONESIE EPIPHANY

So your little one's nappy has gone nuclear and poo has made its way up their back. You want to change them, but the idea of pulling a soiled onesie over their head fills you with horror.

Fear not! Those little shoulder flaps at the top of the garment are not just for decoration, they're designed to allow the head opening to become wide enough for you to slip the onesie down along your baby's body, so you can avoid spreading the mess. Simple, yet a lifesaver, especially when changing a newborn up to ten times a day.

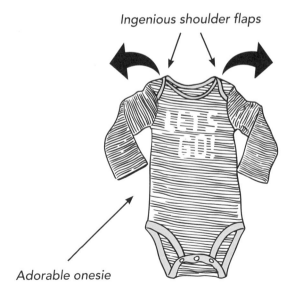

Ingenious shoulder flaps

Adorable onesie

Talking of poo, here's a handy hack for cleaning meconium (newborn baby poo) off your precious little one's behind. Meconium is like Marmite and if you're strictly following the recommendation that for the first six weeks you should use only cotton wool and water on your baby's backside, it can take a while to shift.

But there is a substance that is harmless to newborn skin that will make light work of this job: olive oil. Just rub a small amount of olive oil onto your baby's bottom after each nappy change for the first few days and the meconium will wipe away with ease.*

*Midwife approved.

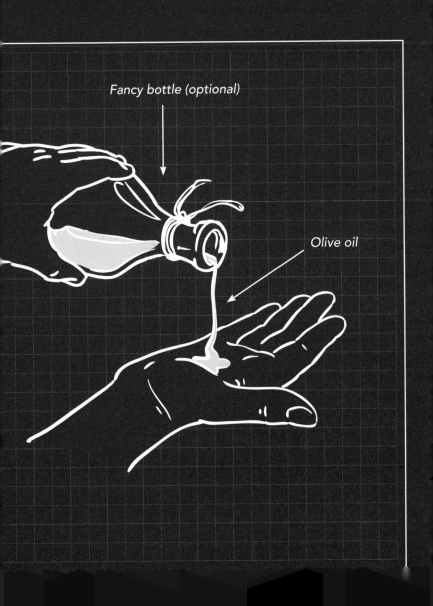

Fancy bottle (optional)

Olive oil

SOCK BUDDY

Babies love contact and cuddles, and while it can be soothing for both of you, it makes it difficult to put them down for a nap. Here's where a couple of common household items come in handy.

Fill a sock with rice – make sure the sock is the extra thick woolly variety so there's no risk of rice grains poking through – and securely sew up the opening. Be sure to give it a shake to make sure none of the contents will escape and place it in the microwave for one minute on maximum power. Let it cool to body temperature (touch it to test) and then lay it beside your baby so they can be soothed by its warmth.*

*Your rice sock should last for a while but exercise common sense – if it starts to smell or the stitching comes loose, throw it away and make a new one.

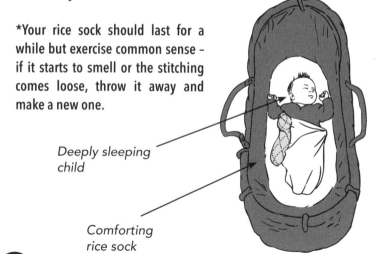

Deeply sleeping child

Comforting rice sock

COT CHANGE

This is a common-sense hack for changing your baby's bedding in the night – an unavoidably frequent occurrence with a newborn.

It can be such a performance finding replacement sheets and mattress liners while you're half asleep and trying to soothe a baby at the same time. The simple answer is to double up on the bedding. After you have placed a mattress liner and a sheet down, place another mattress liner and sheet on top, so you only have to remove the top two layers, causing minimum disruption.

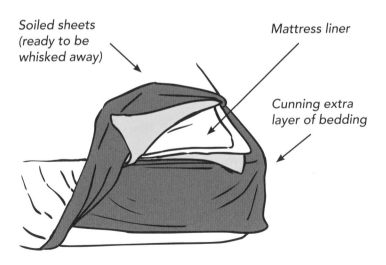

Soiled sheets (ready to be whisked away)

Mattress liner

Cunning extra layer of bedding

CRADLE CAP
COCONUT REMEDY

Babies are prone to dry skin, particularly 'cradle cap'. This layer of crusty skin on the top of their head can stick around for a long time.

Rather than heading to the nearest chemist, try using a little coconut oil on the affected area. Not only is it harmless if ingested, but it's gentle on your baby's skin. Coconut oil can also be used to soothe insect bites, eczema and nappy rash – and it smells good too!

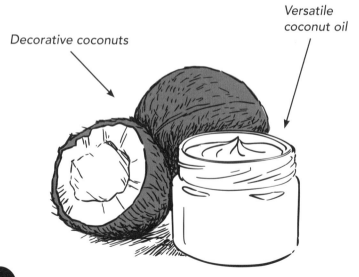

Versatile coconut oil

Decorative coconuts

STAIN PROTECTION

How many times have you got ready to go out, only for your small person to unexpectedly accessorise your outfit with vomit or poo? It's an occupational hazard when you're a new parent.

So try this hack to save your clothes. Get yourself a lab coat – yes, those things you used to wear when performing experiments in chemistry at school. That way, you can keep your clothes immaculate, rather than being known as the crustiest parent in town.

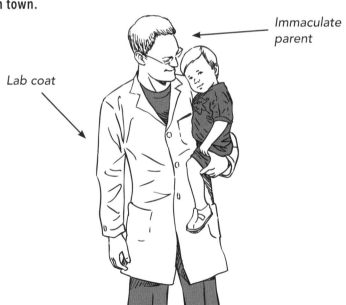

Immaculate parent

Lab coat

TEETHING RELIEF

Teething is an uncomfortable experience for both children and their parents. Here's a simple hack to soothe sore gums. Make sure your hands are clean for this one.

When your baby is teething, all they want to do is chew! Allow your little one to chew your fingers. Another simple thing to do is rub their gums with your finger, as the pressure will soothe them.

Teething child

Your fingers

BOOB CUBES

Freezing breast milk in an ice-cube tray – preferably one with a lid – makes it easier to defrost quickly and the cubes are ideal for mixing with baby food when weaning. Also, in a standard ice-cube tray, each cube amounts to one ounce (28 ml) of milk, so you know how much your child is having.*

*Defrost breast milk in the fridge. Once defrosted, it can be stored in the fridge for 12 hours. Do not refreeze breast milk once it has thawed.

Do not defrost or heat breast milk in a microwave. If time is of the essence, place the container of breast milk under cool, then warm running water, or in a bowl of warm water. Dry off the outside of the container before you open it and use immediately.

Do not repeat the words 'boob cubes' in public places.

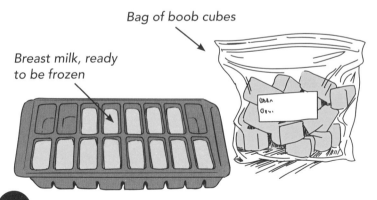

Bag of boob cubes

Breast milk, ready to be frozen

MILK ON TAP

When you're out and about, it can be tricky to find a place to heat up your bottle of milk in a hurry. It normally involves going to a cafe and asking nicely - but what if you're in the middle of nowhere? This simple hack does the job.

Place a bottle of breast milk in a travel mug and add hot water to the mug to keep it warm, so it's ready to use as soon as you need it.*

*It's recommended that you should discard heated breast milk after two hours.

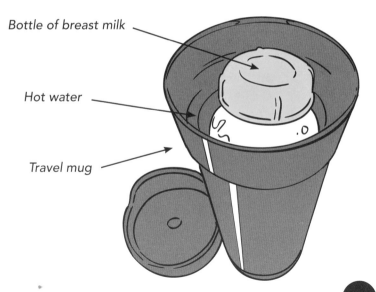

Bottle of breast milk

Hot water

Travel mug

HOME SAFETY HACKS

Child safety in the home is paramount once your little one starts to roll, crawl and toddle like an excited meerkat. These hacks offer low-cost and ingenious ways to eliminate potential dangers in the home.

NON-SLIP SLIPPERS

Once your child starts to toddle you need to buy supportive footwear to save them from accidental tumbles, bumps and scrapes. While your toddler might love wearing them, slippers – especially if you have wood or tile floors at home – can be precarious. This hack eradicates the need to fork out for expensive slippers with grippers.

Make your own simply by taking an existing pair of slippers and applying blobs of glue onto the soles with a hot glue gun.* Once dry, the blobs of glue do exactly the same job as grippers.

*These can be picked up cheaply at hobby and craft shops.

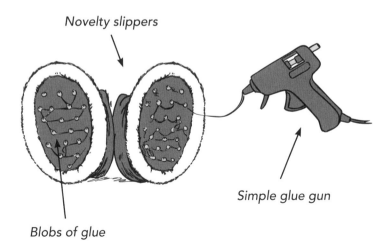

Novelty slippers

Simple glue gun

Blobs of glue

BEDTIME BUMPER

This hack will prevent many a bump in the night for anxious parents whose kids seem to involuntarily fall out of bed, having been previously used to their sturdy, four-sided cot.

Get hold of a pool noodle (that's the long cylindrical float used to aid swimming) and place it under a fitted sheet on the side of the mattress open to the room, creating a soft barrier. The noodle will help prevent your child from rolling over and out of the bed in the middle of the night.

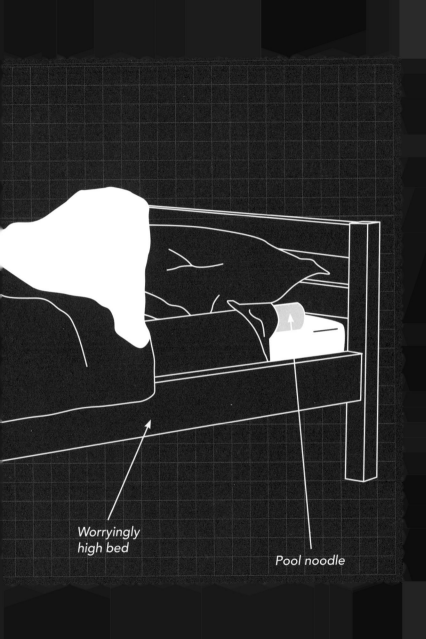

Worryingly
high bed

Pool noodle

FINGER-SAVING DOOR STOPPER

If you're worried about your child pinching their fingers in the door or they have a habit of slamming doors, then this hack is for you.

Cut a 15-cm section from a pool noodle (if you bought extra-long ones for the Bedtime Bumper hack, then you're winning!) and slice it lengthways on one side, like you would a baguette. Slip the section of pool noodle over the upper edge of the door, and you have a quiet door stopper and no more trapped fingers.

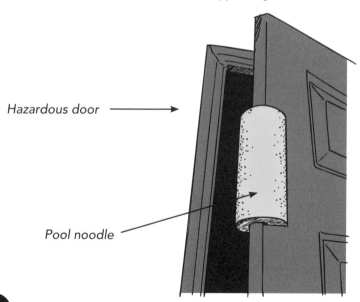

Hazardous door ⟶

Pool noodle

BOOKSHELF SAVER

Toddlers love to explore and there are only so many things that you can place out of their reach. This hack will mean you can keep your books on the shelves without the worry of your child getting injured or your books getting damaged.

Use inner tubes of bicycle tyres (make sure they're clean!) and stretch them over the entire shelf - front to back - that way, when your child makes a play for your books they will stay firmly in place.

Weighty tomes

Curious child

Inner tubes of bicycle tyre

PLUG AWAY

Some children have an almost magnetic attraction to all things electrical, particularly wires, plugs, multi-plugs, extension leads and surge protectors.

To prevent your child from exploring potentially hazardous areas involving plugs, etc., get a large plastic container with a lid and cut a hole in the side (two if needed). Then, place the surge protector in the box and thread the wires and cords through the hole. Now seal the box with the lid to keep the electrics out of reach.

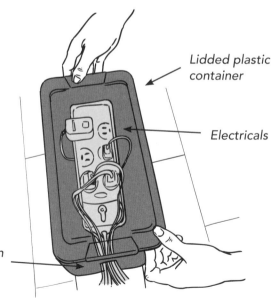

Lidded plastic container

Electricals

Hole cut in the side

WINDOW STICKERS

Large windows and glass garden doors are a head-bumping hazard for small children, so here's a simple hack to make glass more visible.

Use brightly coloured vinyl stickers with adhesive backing (otherwise known as decals, which won't leave a mark) to decorate the window – if you have bigger children you can make this into a fun activity for them. The window or glass door will now be easier to see, preventing your baby from crawling head first into it.

Large glass door

Brightly coloured vinyl stickers

RUBBER BAND LATCH-CATCH

If you have experienced the worry caused by a faulty latch on a door (especially if it sticks and traps your toddler in the bathroom) then you'll be pleased with this hack.

Loop a strong rubber band around the base of the handle on one side of the faulty door, twist it over and then loop it around the base of the handle on the other side. The rubber band will make a latch-catching barrier where it forms a cross, which will hold your faulty latch back, avoiding any sticking-door scenarios.

Door handle

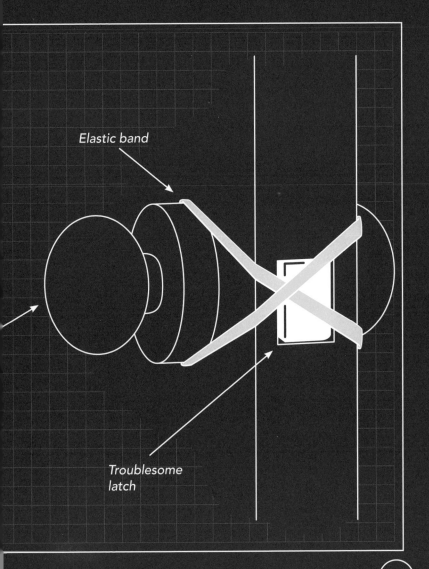

Elastic band

Troublesome
latch

TENNIS BALL
TABLE CORNERS

Coffee and dining tables become highly dangerous once your child is mobile - especially the corners. If you don't want to pay for expensive plastic edges (which aren't even cushioned) then tennis balls are the answer. Take as many tennis balls as you require and cut a thin wedge out of each before slipping them onto the corners. It might look like you're preparing for a special tennis-themed dinner party for a while, but that's a small price to pay for child safety!

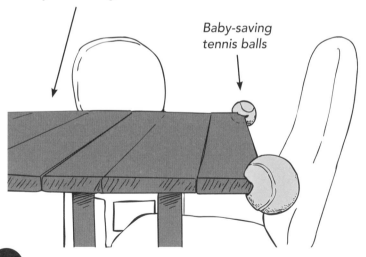

Shabby-chic dining table

*Baby-saving
tennis balls*

BABY-PROOFING GLASS TABLES

Glass tables may look stylish but, like table corners, they can be harmful to children. Here's a way to soften those hard edges.

Get a length of foam pipe insulation, available at good hardware stores. Cut to the size of the table edges and cut into it lengthways, so that the foam can be slipped onto the edges. Use duct tape to secure it in place. This safety method can also be used on hearths and any other furniture with pointy edges.

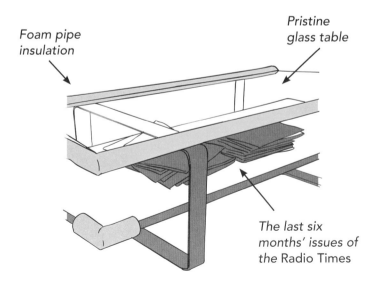

Foam pipe
insulation

Pristine
glass table

The last six
months' issues of
the Radio Times

TRAMPOLINE SAFETY

Trampolines are immensely popular with children, but they also account for a large number of admissions to A & E, not least because their springs are potentially hazardous. Here's one way to make them safer.

Back to our old friend the pool noodle. You will need a fair number, depending on the size of your trampoline (eight to ten should do the trick). Cut them so they're the same length as the trampoline springs, slice them lengthways (as with the Finger-saving Door Stopper) and attach one to each spring. Your child's limbs are no longer at risk of getting trapped.

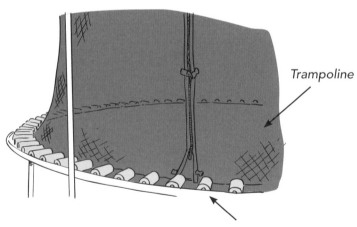

Trampoline

Sections of trusty pool noodle

SPARKLER SAFETY HOLDER

Fireworks are one of the highlights of Bonfire Night for children, but every year there are accidents involving sparklers. This hack will allow your child to play with sparklers and be safe.

All you need is a large plastic cup. Make a hole in the bottom and thread the sparkler through it, being careful that the sparkler end remains on the outside of the cup. Your child can hold the sparkler, but is protected from the heat by the cup. Make sure there is a bucket of water that the sparkler can be dropped into once it's burnt out, and always supervise your child closely while the sparkler is lit.

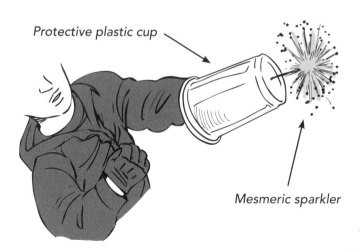

Protective plastic cup

Mesmeric sparkler

NON-SLIP RUG

No matter how much you supervise their playtime, falling over is a given for kids, but you can do your best to avoid it by making potential obstacles on the floor safer.

If you have rugs and doormats that are devoid of a grippy underside, don't splash out on an expensive non-slip mat cut to size - simply apply some lines of acrylic sealant (the kind you use in the bathroom) to the bottom of the mat or rug and it should stay put.

Rubbery, grippy sealant

Now non-slip doormat

FIRST AID HACKS

The health and well-being of your child is not to be taken lightly, and while the GP, midwife or health visitor should be your first port of call in the event of your child becoming poorly, the following pages contain hacks that should make caring for them that little bit easier.

BANANA SKIN
BITE RELIEF

Bites can be excruciatingly itchy, but this hack will alleviate the irritation without using medicated cream – and it will make you see bananas in a whole new light!

The next time your child has a bite, peel a banana and, using the inner side of the skin, rub it on the bite. The banana skin will reduce the swelling and soothe the itch.

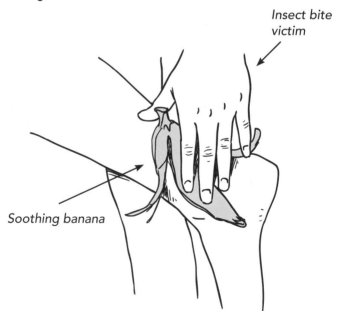

Insect bite victim

Soothing banana

ASSISTED SPLINTER-REMOVAL

Splinters can be painful, but here's a hack to remove them with ease.

Pack a little baking soda into your first aid kit. When that splinter rears its ugly head, wet the offending area with water and sprinkle a little of the baking soda onto it, then carefully cover with a plaster. Leave it for a day or so. When you peel back the plaster the splinter will be raised out of the skin, making it easy to pick out.

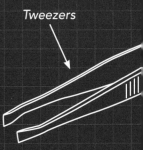

Tweezers

The offending splinter

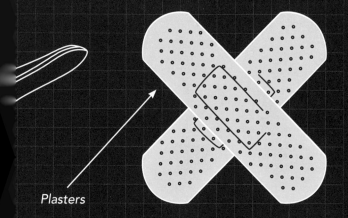

Baking soda

Plasters

PVA SPLINTER-REMOVAL

Here's another hack for removing splinters if you don't happen to have any baking soda in the cupboard.

Pour a small blob of white PVA glue over the splinter. Leave the glue to dry and then carefully peel it off. The splinter – if it's not too deep – will come out attached to the glue.

Splinter site

Dried PVA glue

PVA GLUE

MEDICINE TRACKER

With lack of sleep and the million-and-one things on your to-do list it can be difficult to remember when you last gave your child their medicine, but help is at hand.

The low-tech way is to draw a table with a permanent marker on the back of the bottle, so you can cross off each dose as you go. If you have a smartphone or similar device, there are apps, such as Medisafe, that will send you a reminder when it's time for the next dose.

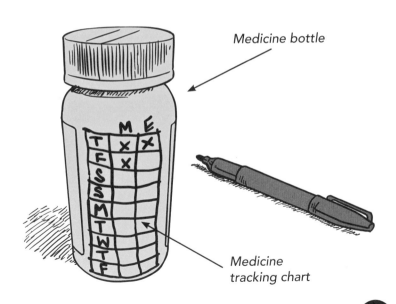

Medicine bottle

Medicine tracking chart

NON-DRIP ICE PACK

Sprained ankles and scraped knees no longer have to be a soggy affair, thanks to this ingenious hack.

To make a non-drip ice pack, soak a sponge in water then place it into a ziplock bag before freezing it. When you come to use it, the melting ice will collect in the bag instead of running all over your child's clothes and the carpet. Keep a few sponges on ice so they're ready when required.

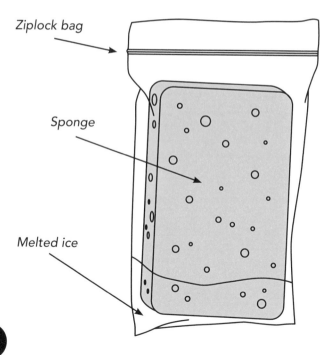

Ziplock bag

Sponge

Melted ice

MARSHMALLOW ICE PACK

If you're looking for a softer and not-too-cold ice pack for little ones, try using marshmallows!

Simply fill a ziplock bag with marshmallows, place in the freezer until your child suffers a bump or graze and then apply to the affected area. The novelty of using marshmallows will help take their mind off their boo-boo too!

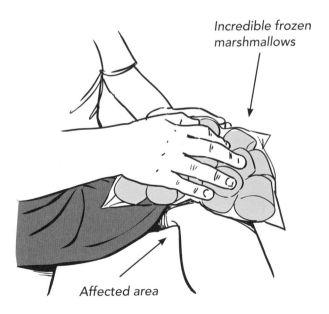

Incredible frozen marshmallows

Affected area

TICKLY THROAT RELIEVER

Itchy throats aren't fun, especially as you can't reach the itch to scratch it – but here's a hack to help.

The next time your child complains of a tickly throat, try touching their ear. As crazy as it sounds, massaging their earlobe between your thumb and index finger will make the itch disappear. Massaging the ear stimulates nerves that can cause a tiny muscle in your throat to spasm, thus 'scratching' the maddening itch.

Hack can also be utilised by mums, as seen here

MARSHMALLOW THROAT SOOTHER

It turns out marshmallows have further health benefits because they can be used to soothe a sore or scratchy throat. This is great news for parents of children who are reluctant to take medicine.

The gelatin in this tasty treat acts in the same way as a throat lozenge. Give your child three or four (or more, if you're feeling generous) to suck on and they will feel better in no time, but watch out for fake sore throats as these are likely once they realise they can have marshmallows!*

*This hack is strictly for children above 36 months as marshmallows are a choking hazard for babies and toddlers. Not suitable for vegetarians or vegans.

Throat-soothing confectionery

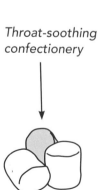

MARSHMALLOWS

FRUITY COUGH SYRUP

A bad cough is no fun for anyone, and it can be distressing for a small child to suffer a coughing fit in the night. This hack will help them feel much better.

The next time your little one is suffering from a cough, offer them some watered-down pineapple juice to drink. It has a similar soothing effect to cough medicine – plus, it counts as one of their five-a-day!*

*Not for children under 12 months.

Believe it or not, this is a pineapple...

... and this is its juice...

SLEEP-EASY SOCKS FOR SNUFFLES

It's tough on little ones when they have a cold, especially if it means broken sleep – which can be tiring for you too! Try this hack the next time your child has a blocked nose.

Apply vapour rub to the soles of their feet and cover with socks – that way, they can't accidentally ingest the rub if they play with their feet in the night, and it means that the vapour rub doesn't stain the bedding.

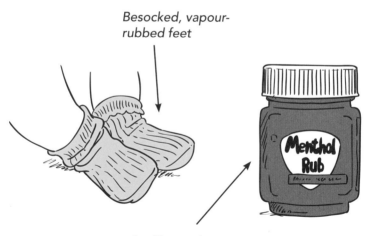

Besocked, vapour-rubbed feet

Snuffle-combating menthol rub

INSTANT SUNBURN RELIEF

No matter how many times you apply sun cream to your kid, there will come a day when you'll miss a spot and inevitably they'll start to complain of hot, itchy skin. Here's a natural sunburn relief hack that's easy to make and can be stored away until it's required.

You will need aloe vera gel and an ice-cube tray – that's it! Squeeze the aloe vera gel into the ice-cube compartments to three-quarters full, add water to fill the compartments to the brim and place in the freezer. Apply to the sunburnt area for instant relief.

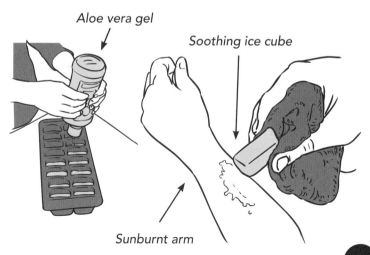

Aloe vera gel

Soothing ice cube

Sunburnt arm

AMAZING GRAPE JUICE

This hack is a life-saver for families who are prone to catching stomach bugs. One way of combatting this is to stay at home and not mix with other babies and children, but this hack is far less antisocial.

The next time a member of your family has tummy trouble, get some 100 per cent red grape juice and make it part of everyone's diet. Here's the science bit: the grape juice alters the pH levels in your intestinal tract stopping the virus in its tracks and flushing it out of your system. The juice also contains anti-viral chemicals, vitamin C and antioxidants.*

*Not for children under 12 months.

Amazing grape juice!

CLEANING HACKS

Children = mess. Here at Life Hacks HQ we take stains seriously. These cleaning hacks will leave your home stain and germ-free (for at least 20 minutes!).

GLITTER BE GONE!

Craft time often means that the glitter comes out, and the stuff gets everywhere – embedded in carpets and coating every surface you can imagine. This simple hack will keep your house glitter free.

Invest in a lint roller specifically for tidy-up time and roll it over glitter to remove every last speck – you can even make it into a fun game for the kids. Happy days!

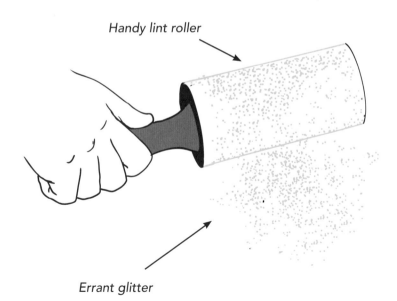

Handy lint roller

Errant glitter

CRAYON REMOVER

Crayon scribbles on the wall are the bane of any household with young children. But this way of removing them (the scribbles, not the kids) will put your mind at ease.

Take a cloth, spray a little water-displacing lubricant on it (the kind that comes in a bold blue and yellow can) and apply to the offending area. The crayon marks will magically disappear!

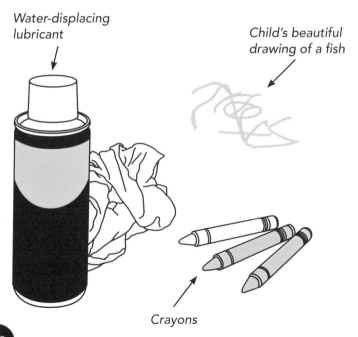

Water-displacing lubricant

Child's beautiful drawing of a fish

Crayons

BREAD ERASER

Question: What do you do if your kids scribble all over your walls (and the Crayon Remover hack on p.54 isn't suitable, because you don't want your living room to smell like a garage)?

Get a slice of white bread (semi-stale works best), remove the crusts and scrunch the soft centre into a ball. Wipe the wall with a soft cloth and then rub your bread ball over the pencil or crayon to erase the offending marks. Yes, you could use an eraser, but scrubbing your walls with a carbohydrate is much more fun.

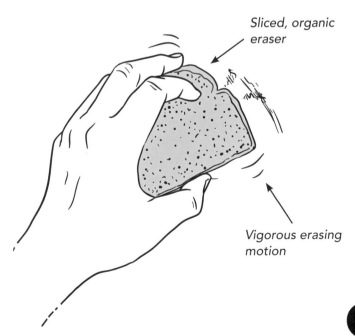

Sliced, organic eraser

Vigorous erasing motion

MAGIC WATER-RING REMOVER

When your children (or friends) are more inclined to use your coasters as mini Frisbees than for safeguarding your table, you're going to have to deal with the threat of ugly water marks. If the worst happens and you end up with a stain, rest assured – your table is not ruined!

Use a hairdryer set on high and hold it close to the water mark. Watch it disappear before your very eyes (but this could take some time, so don't nod off). Then rub a little olive oil into the area to moisturise the wood. Now you can sit back with a cuppa and admire your handiwork (but use a coaster!).

Hairdryer

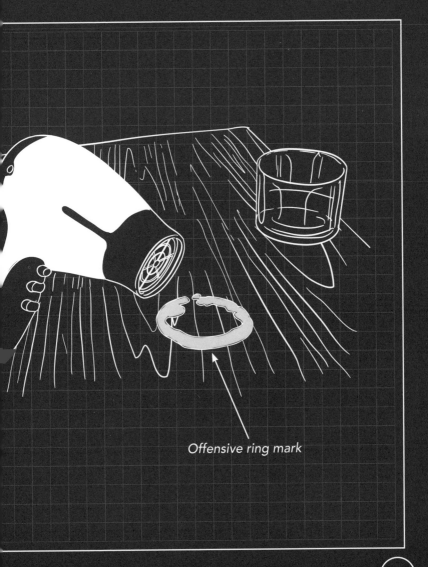

Offensive ring mark

DE-GERM TOYS

Plastic baby toys get handled by sticky fingers and dribbled on constantly, particularly when your little one is teething or just likes to put everything in its mouth. Here's a cheap way to sterilise them.

Instead of using up an industrial supply of anti-bacterial wipes, just put them on a hot wash in the dishwasher every so often. Your baby's toys will come out sparkling and they'll be hygienically clean - unless you have a dirty dishwasher!

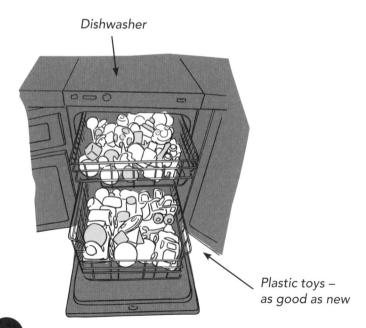

Dishwasher

Plastic toys –
as good as new

BANANA STAIN REMOVER

Until you became a parent you had no idea how difficult it would be to remove banana stains from clothing. When you get a smudge of banana on clothes quick action is necessary before it becomes brown and permanent, so here's what you should do.

The first thing is to scrape off any excess banana with a spoon or knife, then rinse the area with cold water. Next, rub liquid detergent into the area and leave for a few minutes before washing off with hot water through the back of the stain, so it forces the banana stain out. Apply stain-removal spray or gel and then wash in your washing machine as normal. And in future, remember to wear your lab coat (see page 15).

Liquid detergent

The scene of the crime

CHEWING GUM REMOVER

Chewing gum on clothing is a nasty business. You know the scenario; you or your beloved child sits down on the bus or train only to discover that someone has left chewing gum on the seat and now the gum is stuck to you! But wait, there's a hack on the horizon...

Once you're home, remove the gummed-up item of clothing and put it in the freezer. Leave it in there for about an hour or until the chewing gum is rock hard, and then just pick it off.

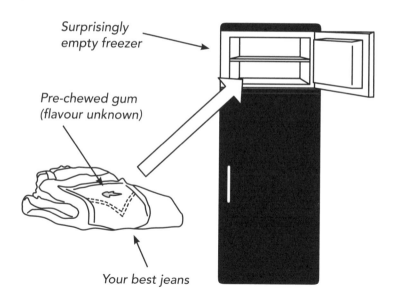

Surprisingly empty freezer

Pre-chewed gum (flavour unknown)

Your best jeans

BABY-POO STAIN REMOVER

Poo-stained onesies are the bane of many a new parent – as with banana stains, a swift response is required. Remove the item of clothing and rinse under warm water to remove as much of the poo as possible. Then place it on a hot wash with a child-friendly stain remover. The poo should disappear, and if it doesn't, try sun-bleaching by hanging it out to dry in direct sunlight.

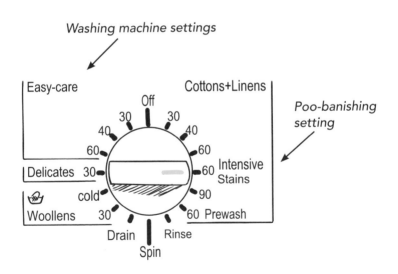

Washing machine settings

Easy-care

Cottons+Linens

Poo-banishing setting

Off

30   30

40   40

60   60

Delicates 30   60 Intensive Stains

cold   90

Woollens 30   60 Prewash

Drain   Rinse

Spin

BERRY STAIN REMOVER

It's great when your child loves soft fruit, but the mess that these juicy treats can make can be off-putting. Don't let it deter you from feeding your child a healthy snack though, because there's a simple hack to remove the stain – and you don't need to act immediately.

First, remove the item of clothing from the child, then scrape off any excess goo from the stain and cover the area with salt. The salt will draw out the juice. Leave for an hour, then rub off the salt before placing the item in the washing machine for a normal wash with some child-friendly stain remover.

Offending berries

Berry stain

Stain-remedying salt

PERMANENT MARKER REMOVER

If, by some cruel twist of fate, your child has managed to get hold of a permanent marker and has 'got creative' with it all over your priceless furniture, fear not.

Go to the bathroom and pick up the toothpaste - use the plain white variety - and spread it over the mark. Leave it for ten minutes, then rub it off with a damp cloth and the permanent ink should disappear. On carpets use white vinegar, and on clothes use hand sanitiser, then buy your future Picasso some chalks or water-soluble pens so it doesn't happen again!

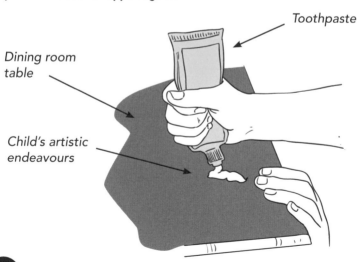

Toothpaste

Dining room table

Child's artistic endeavours

SICK STAIN AND
SMELL REMOVER

Here's what to do when your child is sick on the carpet or furniture (it works for adults too, but they should know better!).

Pour baking soda over the sick until it's completely covered. The baking soda will absorb the sick, including the smell, which can be vacuumed up and disposed of. If a stain remains, try using a little more baking powder and white vinegar, which will react with the baking soda to draw out the stain.

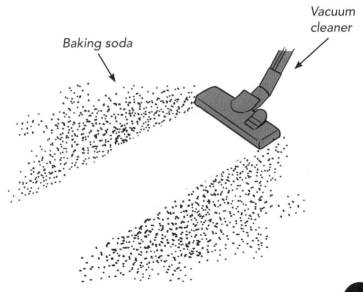

Vacuum cleaner

Baking soda

PEE STAIN AND
SMELL REMOVER

Little accidents in the night can be frequent when your child is potty training. Here's a simple and safe hack for removing pee stains and smells.

Mix 230 ml of hydrogen peroxide (three per cent), which can be purchased at a chemist, three tablespoons of baking soda and a couple of drops of washing-up liquid. Place the solution in a spray bottle, give it a shake and use immediately, spraying over the stain. The stain will disappear within ten minutes and any residue left by the baking soda can be vacuumed or brushed off. Discard any leftover solution as the active ingredients cease to work after a short time.

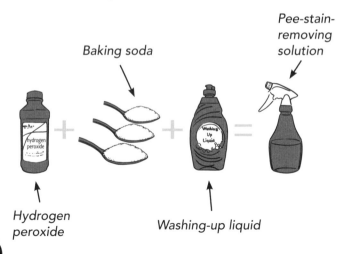

Pee-stain-removing solution

Baking soda

Hydrogen peroxide

Washing-up liquid

BABY SOCKS SAVER

Socks always seem to disappear in the wash – it's one of the great mysteries of life. So what hope is there for your baby's socks and small items?

To stop this nonsense, invest in a mesh bag. Pop your baby's socks inside, place the bag in the washing machine and, as if by magic, the same number of socks will be there when you take the bag out again. Better still, give everyone in your family their own mesh laundry bag – no more sock sorting.

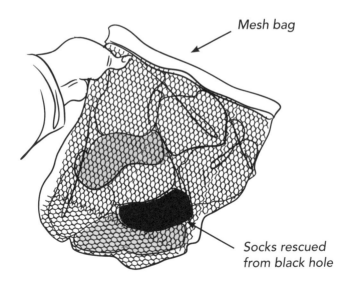

Mesh bag

Socks rescued from black hole

GRASS STAIN REMOVER

If you're the parent of an adventurous child, you will be familiar with grass stains. Here's a handy way of tackling them.

Take one cup of water, one cup of ammonia (you can find this in a hardware or DIY store), one cup of washing detergent and one cup of white vinegar. Put it all in an empty, clean spray bottle and shake. Just spray it on the offending mark and rub it right off. Wash the clothing in the normal way to remove all ammonia from your child's clothing before they wear it again.

Spray bottle

Grass-stained trousers

TIDYING HACKS

If your house has become overrun with toys and you can't even take a bath without several plastic creatures crowding in to join you, or you've tripped over your child's discarded wellies once too often, then this section is for you.

VELCRO TOY STORAGE

Picture the scene: panda, bunny, bear and mutant caterpillar have now reached an age where they are still loveable but perhaps just tend to take up room, getting dusty on the bed or on the floor. Your child won't let you throw them away, so you need a way of keeping them tidy. Is it really possible, you ask? Yes, it is, and this hack will be a lifesaver in the children's bedroom.

Stick a piece of Velcro to the wall (make sure it is the rougher 'hooks' side) and then stick on the stuffed toys. Suspended quite comfortably on the wall, panda, bunny, bear and mutant caterpillar will actually look tidy… unbelievable.

Favourite toys

Velcro strip

TOY TIDY #1

Does it look like a toy bomb has gone off in your child's room? Has it reached a point where there are so many cuddlies in the bed that there's no room for your child to sleep? Discourage your child from being a future hoarder and tidy those toys with these simple steps.

Get a beanbag cover - or make one if you're feeling crafty - and at the end of each day fill the cover with the cuddly toys (soft toys only) and zip it up. Now you have a tidy room and a squishy beanbag to sit on to read the bedtime story.

Large beanbag cover

Beloved cuddlies

TOY TIDY #2

This is a similar idea to the toy-tidy beanbag on page 72, but a little cheaper and classier.

Take a large cushion cover, cut a square out of the front panel and replace it with a thin piece of voile or net curtain fabric. Then stuff the cushion with your child's favourite toys and it becomes portable toy storage that they can take with them on car journeys and holidays.

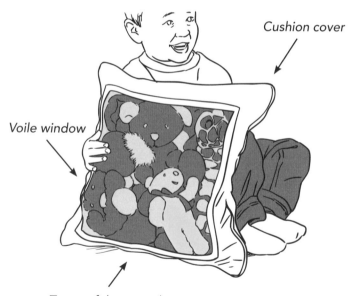

Cushion cover

Voile window

Toys, safely stowed

TOY TIDY #3

Here's another way to keep children's toys from turning the floors of your home into an assault course.

Purchase some hanging baskets and attach them to the wall so they're low enough for your child to help tidy away toys into them, and make a game of filling the baskets each night.

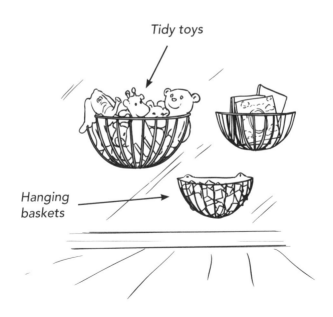

Tidy toys

Hanging baskets

TOY TIDY #4

Here's a brilliant way to store toys when you're tight on space.

Simply attach a hanging shoe tidy to the back of your child's door and fill it up with their toys. Again, you can make a game of putting the toys 'to bed' in the shoe tidy, especially if they are cuddly toys, which look really cute with their heads poking out!

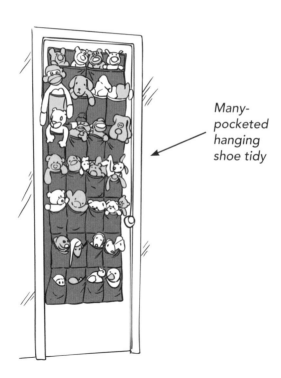

Many-pocketed hanging shoe tidy

TOY TIDY #5

If your toy-storage woes are persisting after all of the afore-mentioned hacks, you might want to seriously consider donating some toys to a good cause – are you trying to start a toy museum?! However, if you're still looking for solutions, try this one.

Apply hooks to your child's wall and hang small washing baskets from them for further storage. Simply unhook the basket when your child wants to play with the toys.

Novelty hooks

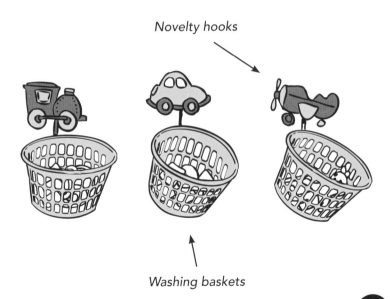

Washing baskets

TOY TIDY #6

This hack is possibly a little more expensive than the previous toy-storage solutions, but it has the extra wow factor of magnets!

To help keep your child's toy car collection organised (and save your feet some drama) try this. Apply a couple of magnetic knife racks to the wall and hang the cars from them – it makes a great display too.

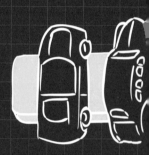

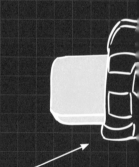

Toy cars

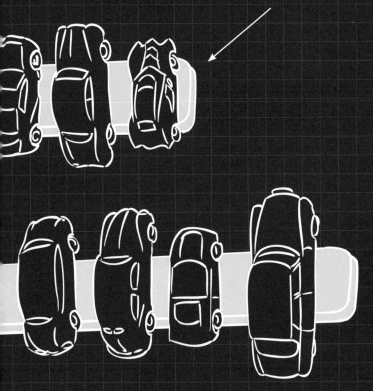

Magnetic knife racks

SHOE RACK HACK

Children have a habit of kicking off their shoes when they enter the house and not putting them away. To prevent these trip hazards and to teach your children good habits, try this hack.

Simply mount some coat hooks in the hallway, low to the floor, and teach your children to hang up their shoes when they take them off. Easy!

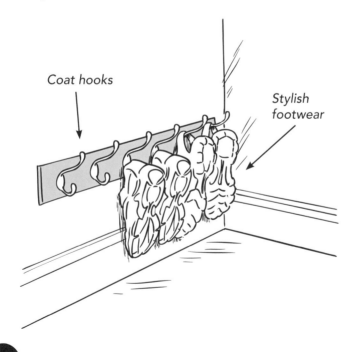

Coat hooks

Stylish footwear

FUNKY SHOE STORAGE

Here's another shoe-storage hack, for the more adventurous homeowners out there.

For a statement shoe tidy that your children will enjoy using, place a wooden pallet on its side, attach it securely to the wall and then invite your children to post their shoes through the holes! Each child could have their own rack depending on their height, so the tallest can have the top slot, etc.

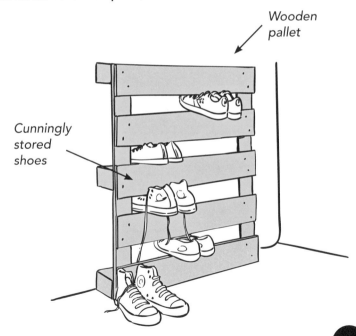

Wooden pallet

Cunningly stored shoes

WINE STORE TO SHOE STORE

Wine racks - the large wooden variety with pigeon holes - make great storage solutions for small items that can easily get lost, such as children's shoes, hats, gloves and swimming goggles, so here's the Life Hacks way to use one.

Place the rack in your hallway and encourage your children to 'post' their items in the slots for safe-keeping and easy access. Just remember it's for the kids and not your wine!

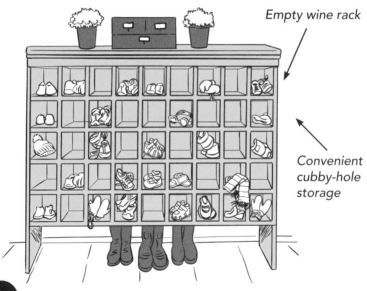

Empty wine rack

Convenient cubby-hole storage

SMART CLOTHES STORAGE

It's good to encourage your children to pick out their own clothes in the mornings, but not so much fun when they empty out the entire contents of their chest of drawers into a large heap on the floor just to find their favourite orange T-shirt. This organisational hack will prevent you from spending hours folding and tidying, and make light work of finding their favourite top.

Arrange folded clothes within their drawers so that they are upright instead of layered on top of each other, so all items can be seen at a glance. It's easy when you know how!

Well-crafted chest of drawers

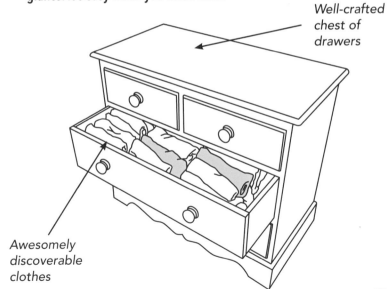

Awesomely discoverable clothes

PENCIL HOLDERS

Here's a low-cost hack to keep your child's colouring pencils organised and in one place.

Store them in milk containers! Wash out the containers, cut out the lower quarters and attach to the wall above their desk with a screw or a strong adhesive.

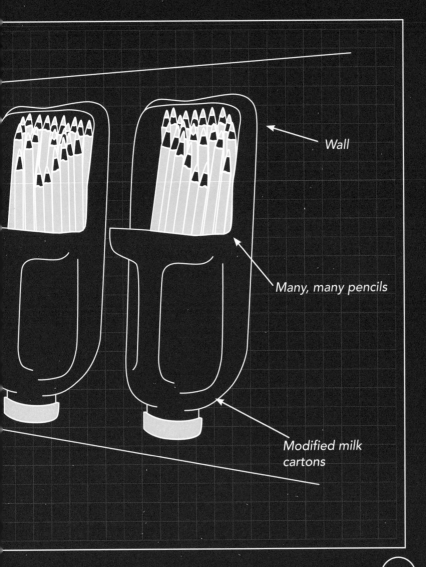

Wall

Many, many pencils

Modified milk cartons

ARTWORK SAVERS

Children seem to produce an enormous amount of 'artwork', but where can you put it all once the fridge door has been liberally covered? Here's how to store away your child's masterworks without them becoming crumpled and damaged.

Simply cut a spent toilet roll tube lengthways and wrap it around the rolled up artwork. This keeps the paper neatly in place and wrinkle-free. Then place it somewhere safe, like in a large plastic container with a lid.

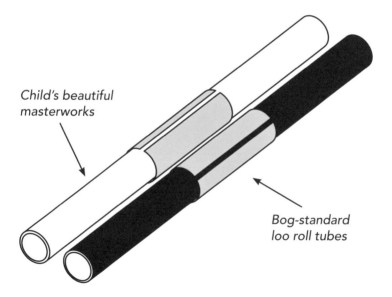

Child's beautiful masterworks

Bog-standard loo roll tubes

BATH TOY ORGANISER

When your little ones start to accumulate bath toys, it can be difficult to know where to store them once bath time is over. Placing them strategically on the corners of the bath means there's no room for the shampoo, and it can start to look a bit untidy. This simple hack will keep them tidied away and accessible for your child to grab when they next have a splash.

Use a rod above the long side of the bath, attaching it securely to the wall, then fit plastic baskets to the rod with shower hooks and fill the baskets with your child's bath toys. This way of storing the bath toys will also ensure that they air dry and don't go mouldy.

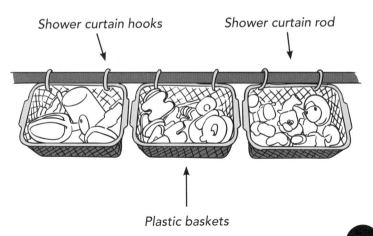

Shower curtain hooks

Shower curtain rod

Plastic baskets

WRAP-UP HACK

Wrapping paper becomes a regular purchase when you have children, what with the endless children's parties and christenings that punctuate your calendar. How often have you searched for some wrapping paper only to find it crumpled and unusable? This hack will ensure your wrapping paper is in tip-top condition whenever you need it.

Secure your rolls of wrapping paper in a clothing bag and hang it in a wardrobe or cupboard until you need it, keeping it tidied away until it's required.

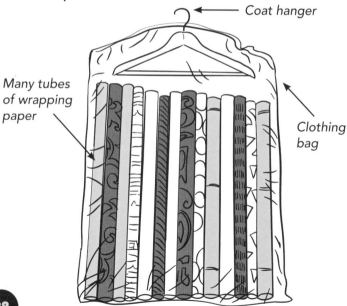

Coat hanger

Many tubes of wrapping paper

Clothing bag

WARDROBE EXPANSION KIT

When you have kids, wardrobe space is hard to come by and it can be tempting to fold your child's best clothes into drawers rather than wrestling with their overstuffed wardrobe. But wait: this amazingly simple hack doubles the capacity of your wardrobe.

Save the ring pulls from cans of soft drinks and beer and link them onto the hook of a hanger, letting them rest at the base where the hook meets the hanger itself. You have now created an extra ring to attach another hanger, amazing!

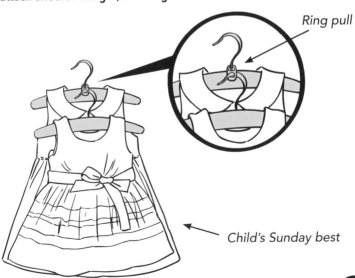

Ring pull

Child's Sunday best

SHOEBOX STORAGE

Here's a storage no-brainer. Instead of throwing away your old shoeboxes, use them as space dividers in your child's clothing drawers.

Paint the boxes, or cover them in bright wrapping paper and fill: sort out your child's knick-knacks, pants and socks, T-shirts, shorts, etc. Put the lids back on and slide the drawer closed. Being tidy has never been so simple.

Repurposed shoeboxes

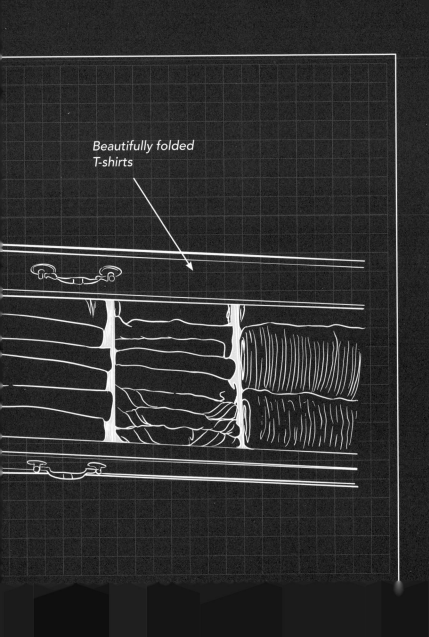

Beautifully folded
T-shirts

HAIR-TIE ORGANISER

If you have a child with long hair, at some point you are bound to find errant hair-ties lurking on every surface in the bedroom. Hundreds of the things come in a pack and within a few hours, without a doubt, they become strewn across the bedroom, some never to be seen again. But Life Hacks has a solution.

Go to your local outdoors shop and purchase – cheaply, of course – a carabiner (the big metal clasps usually used for mountain-climbing rope). Gather up all of the stray hair-ties and click the carabiner closed; now they're all in one place and easy to access ready for the morning rush.

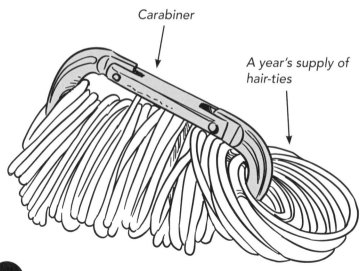

Carabiner

A year's supply of hair-ties

STICK 'EM UP REMOTES

Do you often struggle to find the TV remote and games controllers? This hack will make searching for them a thing of the past.

Nominate an area (the side of the TV cabinet or somewhere of equal surface area) to keep the games controllers and TV remotes, then use Velcro to fix them in place. All you have to do is remember to stick them back (unless you have a child with an obsession for hiding the remote – in which case, you're on your own!).

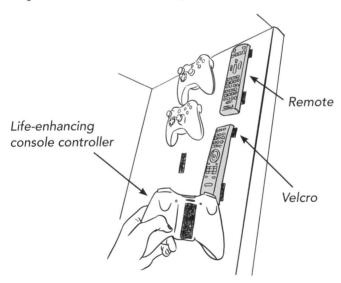

Remote

Life-enhancing console controller

Velcro

FOOD AND DRINK HACKS

This tasty collection of hacks will ensure that your baby is never short of a bib when they need one, your fizzy drinks will stay fizzy and ice lollies will no longer mean sticky hands. There are also some fiendishly clever ways to encourage your children to eat their five-a-day.

BIBS AT YOUR FINGERTIPS

Why is it that you can never find a clean bib when you need one? This hack is the answer.

With just a simple hook attached to the back of the high chair you can have your bibs stored and ready to go. Get yourself a couple of hooks with sticky fixers and you'll never be short of a bib at mealtimes again.

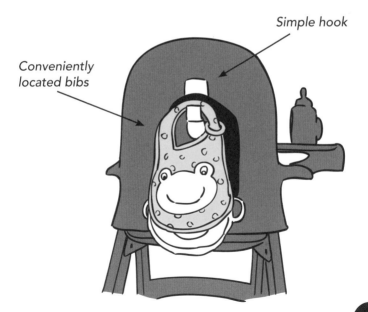

Simple hook

Conveniently located bibs

BABY-FOOD CUBES

Weaning is made simpler and produces less waste with this hack.

Babies eat tiny portions of food when they start on solids, so instead of making a fresh batch every mealtime only to throw most of it away, make batches of pureed fruit and vegetables and freeze the contents in ice-cube trays. Once frozen, transfer the cubes into ziplock bags (write the date on which you froze them onto the bag) and defrost one or two cubes when required.*

*Discard the frozen baby food after one month.

Well-mashed broccoli

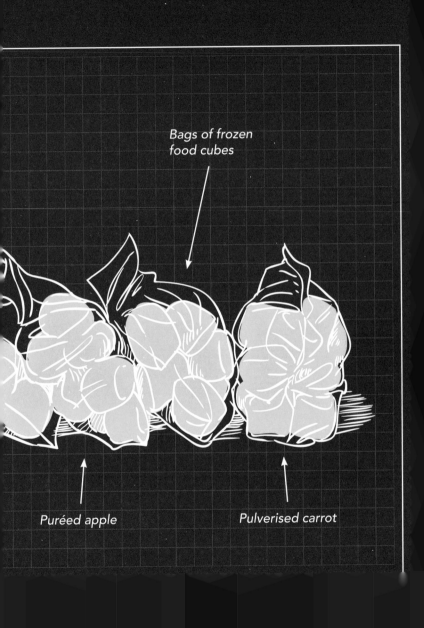

Bags of frozen food cubes

Puréed apple

Pulverised carrot

ICE-CUBE TRAY SNACK CADDY

Do your kids look at a lovingly prepared plate of fruit and vegetables as if you're trying to poison them? If so, this hack could be the answer to your prayers.

Try placing a variety of different bite-sized morsels in an ice-cube tray and make choosing which piece to eat into a game. This magically makes healthy snacks more appealing and it's perfect for the pernickety child who likes to keep their snacks separate.

'Groovy' grapes   *'Brilliant' blueberries*   *Probably best not to eat that*

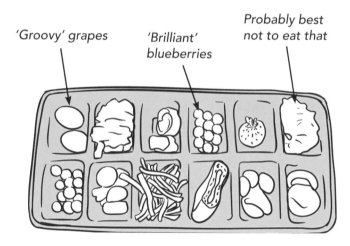

BOTTLE-TOP FOOD PRESERVER

When you need to keep an opened packet of foodstuff airtight, such as dry snacks or pasta, this hack is invaluable.

Cut off the top third of a small plastic drinks bottle to create a 'collar'. With the lid removed, push the top of the open packet through the neck of the bottle. (Pay attention, here comes the good bit.) Fold the end of the packet back over the edge of the neck and replace the lid to create an airtight seal. Genius or what?!

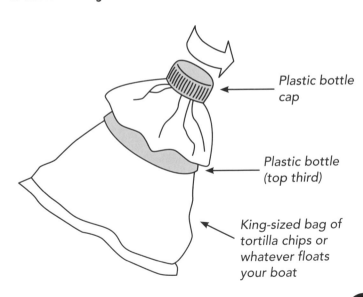

Plastic bottle cap

Plastic bottle (top third)

King-sized bag of tortilla chips or whatever floats your boat

SNEAKY SMOOTHIES

It can be a chore to get your child to consume their five-a-day, but here's the sneaky Life Hacks way to do it.

Prepare a super-healthy smoothie – add whatever fruit or vegetables you have in the fridge or fruit bowl. Then sprinkle hundreds and thousands on top and add a couple of straws so it looks like the most delicious drink ever. Your child will not be able to resist!

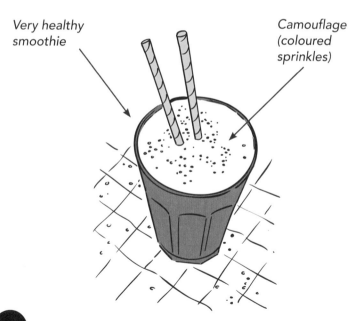

Very healthy smoothie

Camouflage (coloured sprinkles)

APPLE-SLICE TRICK

It's amazing how much better an apple tastes to a child when it's sliced! Here's how to utilise this fact to ensure that your children eat their five-a-day.

Slice an apple into segments and secure the slices together with a rubber band. This will stop them from turning brown. (No one likes brown apples.) Toss this into your kid's lunchbox or take it with you when you're out and about. Whip off the rubber band and you'll have fresh apple slices to hand. Works well for big kids, too.

Ready-sliced apple

Rubber band (not edible)

GRAPE CUBES

To chill a drink instantly on a hot day without watering it down with ice cubes, freeze some grapes and drop them into your glass. When the grapes have thawed, they won't melt and water down your drink.*

*If giving grapes to a child under five years, cut them in half lengthways (top to bottom) as they can be a choking hazard.

Perfectly chilled orange squash

Frozen grapes

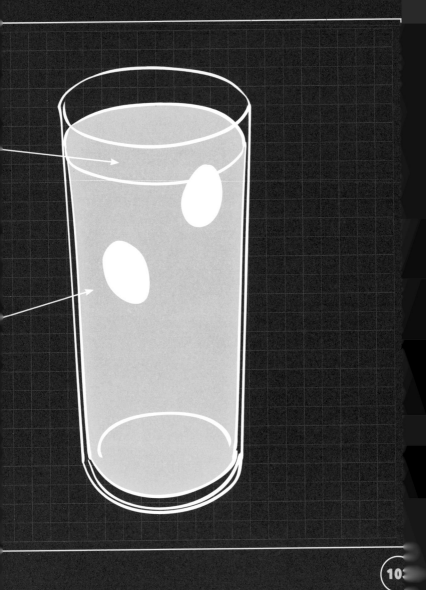

STAY FIZZY

Flat fizzy drinks spell disaster at children's parties. Here's a way to prolong the bubbly soda bliss, presuming you have opted for the plastic bottle version of your favourite drink.

Simply squeeze the air out of your bottle (once you've had a few gulps), so that the liquid is near the top before replacing the lid. This gives the 'fizz' gases nowhere to go and thus keeps your drink invigorated. Works for small or large bottles – not so much for glass ones.

Carbonated (that's posh for 'fizzy') drink

FIZZ

PORTION SIZE CALCULATOR

One of the struggles when feeding children is working out how much to give them. It can be worrying when they say they're full and their plate still has plenty of food on it. Here's a simple hack that gives you a good idea of how much food they need at mealtimes.

Believe it or not, your stomach is roughly the size of your clenched fist – compare that to the size of that takeaway you wolfed down last night! Ask your child to clench their fist and you will see that they don't need a vast portion of food to feel full. Try it at your next meal, better still, if your child is over three years, let them serve themselves.*

*Supervise this and be extra careful if the food is hot.

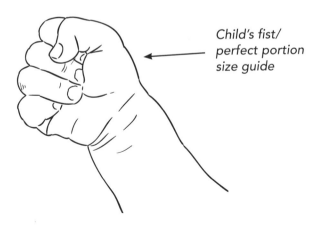

Child's fist/ perfect portion size guide

STRAWBERRY SKEWER

Here's a way to make strawberries even more appealing to your kids!

To hull a strawberry with ease (all of it, including the hard bit under the greenery), simply push a drinks straw from the tip of the strawberry all the way up to the stem. This completely removes the unwanted piece of fruit. Perhaps the clue was there all along… straw-berry.

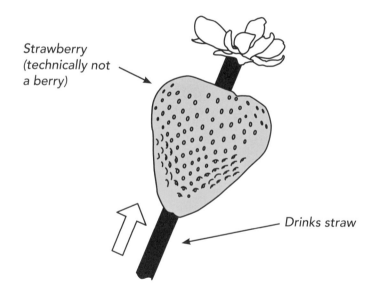

Strawberry (technically not a berry)

Drinks straw

ICE LOLLY DRIP-CATCHER

Dripping ice lollies are a nightmare. The drips get everywhere - on hands, clothes, furniture and pets. Here's how to stop them.

A cupcake case will help you survive this sticky predicament. Just poke the lolly stick down through the middle of the case to create a little cup to catch the offending drips.

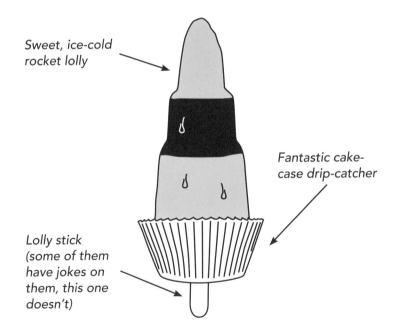

Sweet, ice-cold rocket lolly

Fantastic cake-case drip-catcher

Lolly stick (some of them have jokes on them, this one doesn't)

SOFT DRINK STRAW-HOLDER

Keeping a straw upright in a drinks can is a messy business, especially when the straw bobs up and spills the drink on nice clean clothes. This ingenious idea puts an end to all of that.

Simply turn the ring pull 180° and poke the straw through the hole to hold it in place. So simple! But remember: soft drinks are a treat as they're not good for your teeth.

Securely held straw

Purpose-built straw holder (one free with every can!)

Generic drinks can (other brands are available)

REPURPOSED BAG CLIPS

Here's a hack to stop dry snacks from going stale.

Bag clips are available in shops to stop this from happening, but why spend money when you can use something that costs nothing? Slide the clips off an old trouser hanger and use them to keep staleness at bay. You can now open a fresh bag safe in the knowledge that they don't need to be eaten in one sitting.

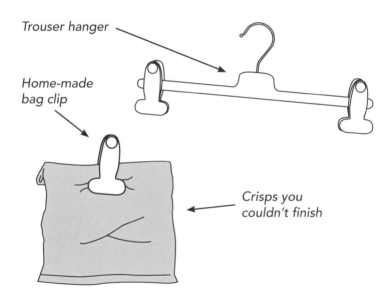

Trouser hanger

Home-made bag clip

Crisps you couldn't finish

HOW TO GET YOUR FUSSY EATER TO EAT

Playing with food is encouraged with this hack as it makes mealtimes fun for fussy eaters.

The next time your little one flatly refuses to take a mouthful, get a dice and ask them to roll it on the table - the number they roll dictates the number of spoonfuls they must eat. Like incurable lottery ticket buyers, your child won't be able to resist the chance of winning big.

Child ready to polish off their greens

Dice

INSTANT
ICE-CREAM SUNDAE

When your child's favourite hazelnut or chocolate spread is almost finished, don't just throw it away – this hack will make an amazing treat out of it.

Fill the almost-empty jar with ice cream and create an instant ice-cream sundae! You can even add chopped fruit, whipped cream and those little sprinkles you put on fairy cakes. If you're feeling generous, donate your delicious Life Hacks treat to your little cherub – or hide in the basement and scoff the lot yourself. Your choice.

Sweet, sweet sundae topping

Delicious chocolate spread

SQUEEZY JUICE

Juice cartons are really handy, but it can be very tempting for little hands to squeeze the box too much, resulting in a juice fountain that invariably stains clothing and soft furnishings. This hack will reduce the risk.

Simply pull up the flaps on either side of the box and get your child to hold them when lifting the box to drink from it. This ensures that the juice ends up in your child's mouth and not on the floor.

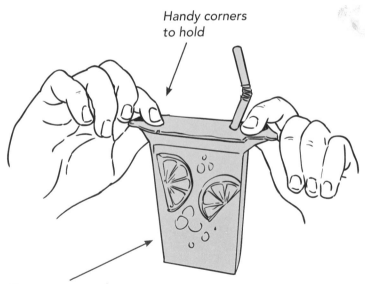

*Handy corners
to hold*

*Common or garden
juice carton*

BUG-PROOF DRINK COVER

Nobody likes bugs in their beverage and children are no exception. Luckily, there's a simple solution. During a birthday party or picnic you can protect your child's juice from thirsty bugs by cutting a small X into the centre of a cupcake case and poking a straw through. Voila! A bug-proof drink cover! Your kids will get a super-cute drink and you can relax, safe in the knowledge that the bees and bugs will realise they're not invited to the festivities and buzz off elsewhere.

Cupcake case (unused)

Bug-free
beverage

YOGURT LOLLIES

Here's a way to make eating yogurt fun for children (and also ensuring that they get their calcium).

Make a narrow slit into each yogurt lid with a knife, push a lolly stick into it and then place the yogurts (upright!) into your freezer compartment. Once frozen, remove the lid, ease out the frozen yogurt lolly and enjoy!

Lolly stick

Yogurt lolly we made earlier

Yogurt pot

PRACTICAL GADGETRY HACKS

You'll wonder how you ever managed without these ingenious hacks. Sometimes it's the simplest things that can make all the difference – such as a target to aid toilet training or sticky tape inside toenail clippers to catch the stray bits of nail. Prepare to be enlightened.

TOILET TARGET

Toilet training is a messy business, and for little boys (and some big boys too!) getting their aim right can take years to perfect. This hack will upgrade their accuracy and keep your bathroom a little cleaner.

Make peeing into a sport and procure a toilet target* with points gained for hitting the bullseye! Create a chart and when they've reached a certain number of points they earn a treat, such as a ball or a colouring book. Hap-pee days!

*Targets can be purchased online.

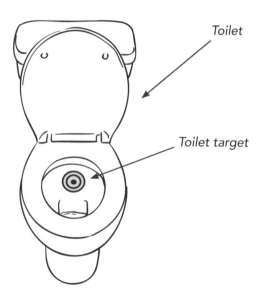

Toilet

Toilet target

TAP EXTENDER

Make your sinks more usable for your child by repurposing a spent shampoo or bubble bath bottle with this hack.

Discard the lid and rinse the bottle to remove any remaining product, then cut a hole in the base of the bottle so it's big enough to fit over the end of the tap. The running water is now within easy reach for little hands, so you don't need to hold them over the sink any more!

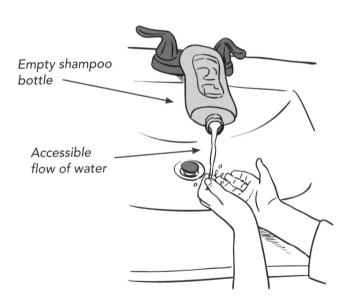

Empty shampoo bottle

Accessible flow of water

TOILET PAPER SAVER

Do your children have a habit of using too much toilet paper when they go to the loo? Here is a very simple hack to stop this from happening.

Attach a marker onto the wall below the toilet-roll dispenser and tell your children that they mustn't go beyond it when pulling off a length of toilet paper. No more blocked toilets!

Precious three-ply quilted loo roll

Toilet roll marker

SIPPY CUP SAVER

Babies and toddlers love watching you picking up after them, especially when they're sitting in their highchair like a pint-sized chieftain, little realising that they could ingest germs as a result of throwing their sippy cup onto the floor (not for one moment suggesting your house is a germ-ridden pit). Outsmart them by using this hack.

Create a tether for their sippy cup with a length of fabric folded and stitched, with one end tied to the sippy cup and the other to a suction cup. Then, simply secure the suction cup to the underside of their highchair tray.

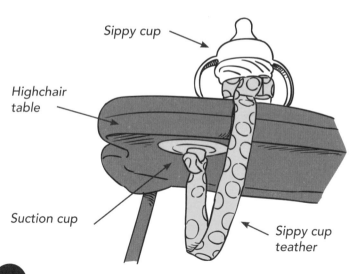

Sippy cup

Highchair table

Suction cup

Sippy cup teather

TIDY TINY NAILS

Toe and fingernail clippings can be sharp, so you don't want them lying around for your baby or small person to accidentally lie on, stand on or eat. This hack ensures that the tiny clippings won't get lost.

Wrap some sticky tape around the clippers and the clippings will become lodged in the tape. This is especially useful once you've realised that the only way to cut your child's nails is while they're asleep, as it eliminates the need to scrabble around trying to pick up clippings without waking them up.

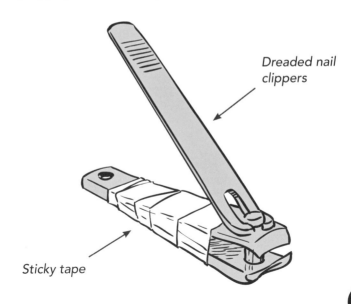

Dreaded nail clippers

Sticky tape

SHOE SCUFF TOUCH-UP

Children's leather shoes get scuffed much faster than they outgrow them - but the greater the scuffing the more fun they're having, right? Regardless, you can help keep their shoes looking something like new with this hack.

Raid your child's art box for a wax crayon that matches the colour of their shoes, warm the crayon slightly in your hands then literally colour in the scuffed areas - job done.

Sturdy leather school shoes

Wax crayon

SMALL ITEM RETRIEVAL SYSTEM

Where do all those bits of Lego, beads and Meccano go when you drop them? I'll tell you where: they're still there, it's just that you're too blind to see them. This hack will get them back.

When you drop something small and can't find it, grab your vacuum cleaner and a pair of old tights. Slip the tights over the vacuum nozzle and fix in place with an elastic band. Run the vacuum over the area where you think you dropped your item and, with a bit of luck, the item will be sucked onto the tights where you can pick it off with ease.

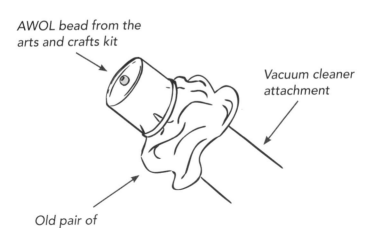

AWOL bead from the arts and crafts kit

Vacuum cleaner attachment

Old pair of tights

SHOE MATCH-UP

The wrong shoe on the wrong foot is not only uncomfortable but can make your child trip up. Here's a way to ensure your child always gets it right.

Get a sticker with a picture on it and cut it in half lengthways. Then attach the left half of the sticker to the inside of the left shoe, so it's against the right-hand edge, then the right half of the sticker to the inside of the right shoe but on the left-hand edge. Then all your child needs to do to get the correct shoe on each foot is to match up the two halves of sticker to make a picture. To kids, shoes might be boring, but stickers are awesome!

Trainers

Large pictorial sticker cut in half

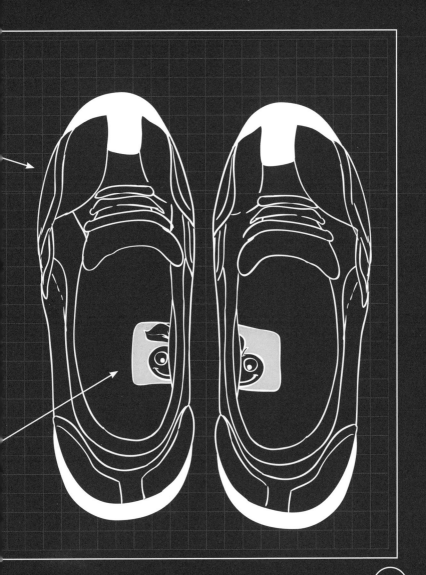

HANDWRITING PRACTICE

Teach your child to hold a pen correctly with the help of a rubber band or a hair-tie.

Put the band or tie around their wrist and twist the band to make a loop and push the pen or pencil through for the child to hold between their thumb and forefinger. This creates the ideal position to hold a pen.

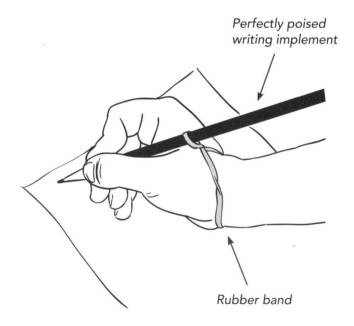

Perfectly poised writing implement

Rubber band

NON-SLIP BOWL

Minimise the risk of bowls or plates slipping on a high chair tray with this oh-so-simple hack.

Wet a paper towel and place it on the tray and place the crockery on top of it; this will stop it from slipping, preventing spills and additional trips to the homeware department for more plates.

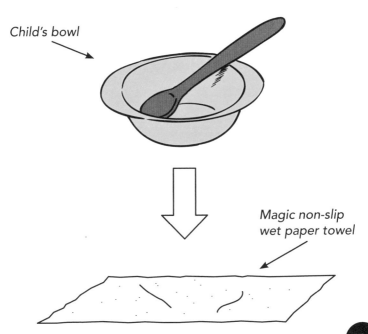

Child's bowl

Magic non-slip wet paper towel

SOCKS FOR FURNITURE

Chances are your children are not as interested as you are when it comes to preserving your wood floors (and who can blame them - they're busy with exciting kids' stuff!). So if you're left cringing every time a chair is dragged carelessly across your pristine floor, this hack will help.

All you need is some pairs of socks, the funkier the better! Pull the socks over the chair legs and not only will your floors be protected from scratches but you'll be spared that awful dragging noise.

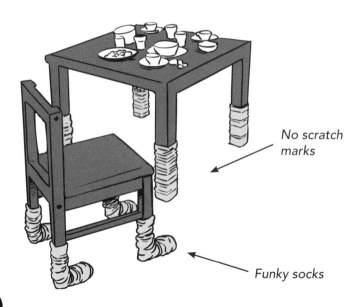

No scratch marks

Funky socks

BABY BATHTUB

Little ones are slippery things, especially in the bath, and when you add toys to the mix things can get even more precarious.

So, what do you need to keep the toys within arm's reach for your child and your child in one place? A plastic laundry basket! That way, your child can sit in it and enjoy bath time without having to scrabble about for toys.* Genius!

*Never leave your child unattended in the bath.

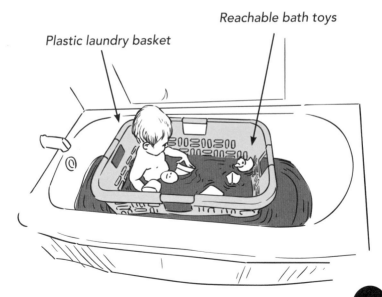

Reachable bath toys

Plastic laundry basket

INSTANT PLAYPEN

When your child starts to crawl or toddle, they can come into contact with all manner of dangers, so what do you do if you need to make a phone call or go to the loo, or even just have five minutes' peace? This hack will solve your problems.

Purchase an inflatable pool - one with high sides - and you have a safe playpen to put your child in. You could even add some balls to create a ball pool - hours of safe fun for your child, and a nice sit down for you!

Hours of safely contained entertainment

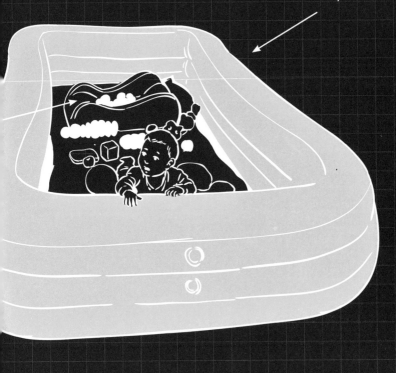

Inflatable pool

THREE-STEP WASH ROUTINE

Help your child to become more independent and learn good habits with their morning and evening wash routine with the help of this simple home-made hack.

Use a small storage bucket and three durable plastic cups to create your child's very own wash station. In the first pot, have their hairbrush and comb, and hair-ties and clips if needed, in cup two keep a rolled-up flannel to wash their face, and in cup three their toothbrush and toothpaste. Number the cups so they know the order in which to perform these tasks, and they're ready for school or bed!

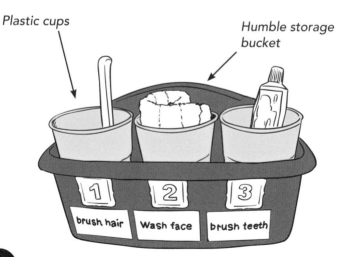

Plastic cups

Humble storage bucket

1 brush hair 2 Wash face 3 brush teeth

OUT AND ABOUT HACKS

Venturing out and about with your children can be a hairy business, but this section offers ways to reduce the stress with some safety advice and a good dollop of common sense.

HOME-MADE SANDPIT

Children love playing with sand, so unless you live near to a beach this hack is for you.

Erect a small tent – preferably not one that you want to use for camping in the near future – and fill the bottom with sand. You now have a sandpit that will neatly contain the sand, shelter your child from the sun and the whole thing can be zipped up at night so the local moggies aren't tempted to use it as a toilet.

Skilfully erected tent

Sand

IDENTITY BRACELET

Heaven forbid that your child gets lost in a crowd, but it's surprisingly common - that's why there's often a lost child point at every festival or organised public event. Here's one way to ensure they have some form of ID on them at all times that will link them to you. Create a bracelet for them with your phone number on it using beads with numbers on, make it obvious that it's a telephone number by adding spaces between the sets of numbers with plain beads. It might come in useful one day.

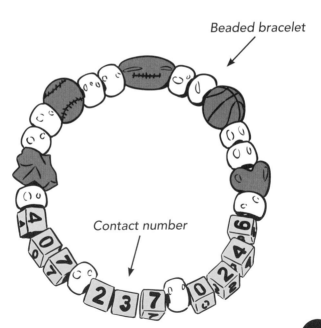

Beaded bracelet

Contact number

CAR SEAT COOLER

Buckles and straps on child car seats can get blisteringly hot if the car is parked in direct sunshine, but this hack will help keep your kids safe.

Keep a spray bottle containing water in the car and spray it on buckles and straps before putting your child in their seat. The water doesn't have to be particularly cold as it will quickly cool down the metal and plastic when it evaporates.

Scorching belt buckle

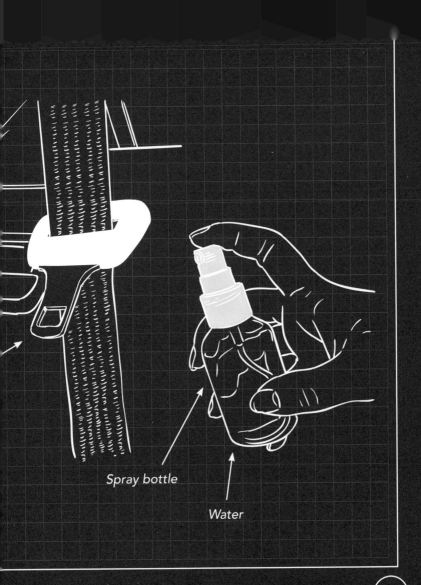

Spray bottle

Water

BEACH SHEET

Trips to the beach are an exciting part of any childhood, but on a sandy beach it's almost guaranteed that grains of sand will get absolutely everywhere. Keep the sand on the beach and out of your sandwiches with this simple hack.

Next time you're planning a beach trip, bring a fitted sheet with you – a double is best, for maximum coverage. Find the perfect spot, stretch out the sheet onto the sand and place items of some weight at the four corners to keep the sheet taut. You now have a sand-free space for your child to play in and eat their sand-free sandwiches.

Heavy items

Fitted sheet

Barrier against marauding sand

CAR PARK SAFETY

Car parks can be dangerous places for small children and it can be easy to get distracted when searching for change for the ticket machine, or when you're filling the boot with your shopping.

Keep your children safe and close to you by attaching stickers to the back of your car - one for each child - and teach them to place their hand on the sticker and stay there until you're ready to strap them into their car seat or take their hand to go shopping.

Sticker

Safely positioned child

IN-CAR TOY TIDY

Here's a great way to keep your child entertained and sustained on long car journeys.

Hook a shoe tidy over the back of the front seat so it's facing your child in the back, and fill it with their favourite toys, a drink, some snacks, wipes and nappies (if required). Instead of them asking you for a drink, snack or toy, they can reach for it themselves! Now that's some smart parenting.

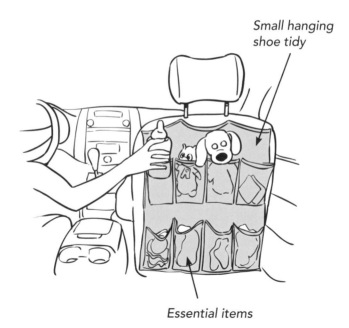

Small hanging shoe tidy

Essential items

IN-FLIGHT SMARTPHONE TV

If you're on a ten-hour flight to Tobago for the family hols and your plane doesn't have seat-back TVs, fret not. There's a simple solution that will keep you and your tribe happy.

Put your phone in a clear plastic bag and attach it to the seat in front of you. Then sit back and relax. Your in-flight viewing system is ready for take-off! Just remember to cut a hole so you can plug some headphones in - not everybody on the plane will want to listen to Igglepiggle.

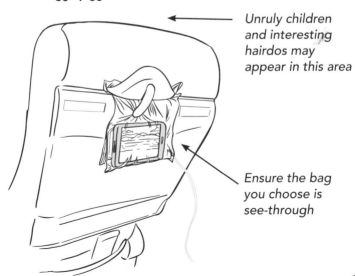

Unruly children and interesting hairdos may appear in this area

Ensure the bag you choose is see-through

BACKPACK BIN LINER

If you go down to the woods for a picnic today, you'd better be prepared with this hack.

One thing your children won't stand for is soggy sandwiches, so use a thick bin bag to line the inside of your bag to keep out heavy downpours and other sandwich-ruining elements. It won't last as long as a professional liner, but it will cost considerably less.

Unreasonably heavy backpack

Super-cheap but effective bin liner

Bone-dry sandwiches and valuables

COLOUR ON THE GO

Arty children will really appreciate this hack. If they like to colour and sketch on train journeys or while out and about, then this is the perfect way to transport their colouring equipment and paper.

Create a colouring kit by repurposing a DVD case. Begin by stripping out the circular DVD holder. Then create a pen-container: cut a piece of card to fit inside the DVD case, and glue onto it a strip of fabric half the size of the card to make a pocket. Then, glue it in place on the right-hand side of the inside of the case. Once dry, fill the pocket with pens and pencils, and slide paper underneath the clip holder on the left-hand side of the DVD case. Decorate the outer case with wrapping paper or artwork.

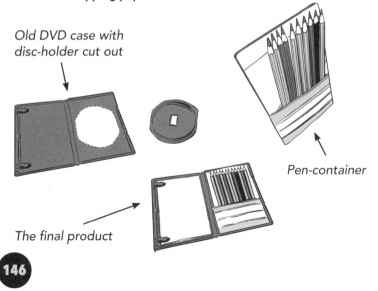

Old DVD case with disc-holder cut out

Pen-container

The final product

BABY SEAT POO PROTECTOR

Babies can make a terrible mess of their car seats when they have unfortunate poo explosions. Here's a way to significantly reduce the clean-up operation that follows.

Cut up a changing-mat liner – the towelling ones that fit over the plastic mat – so it fits snugly into the car seat, then snip a hole for the strap that goes between your child's legs. The next time they have an explosive poo you can simply remove the piece of mat liner without having to deconstruct the entire car seat.

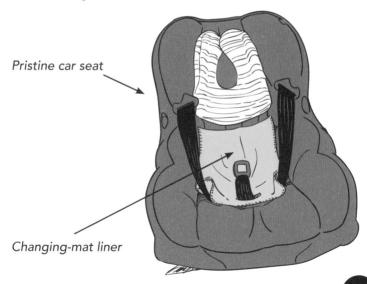

Pristine car seat

Changing-mat liner

ARTS 'N' CRAFTS HACKS

Let your child express themselves with this impressive array of low-cost arts 'n' crafts hacks – just remember to get the overalls on first.

DIY PAINT POTS

Here's a quick and easy hack that will save you money and recycle at the same time!

Create a cheap paint tray with an empty egg box. Fill each compartment with a different paint colour and use cotton-wool buds as paintbrushes. At the end of the painting session, either throw the lot in the bin or place inside a ziplock bag ready for the next session so that the paint doesn't dry out.

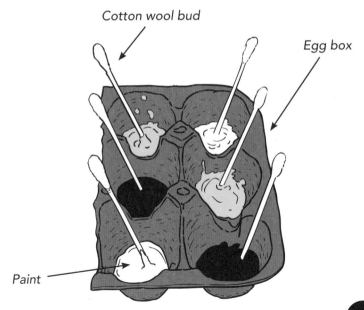

Cotton wool bud

Egg box

Paint

DISHWASHER ART RACK

What do you do when the dishwasher breaks? Well, sparkling clean dishes are out of the question for a while… However, if you are looking to store your child's art equipment, think yourself very lucky indeed.

Use the slots where plates once were to put colouring books and folders of fine art. Use the cutlery pot to store pens, pencils and paintbrushes. Your child will now have all they need to flourish artistically.

Mini-masterpieces

Pens

PVA GLUE SPONGE

Arts and crafts with children can often get very messy. Somehow even when just one gold star needs to be glued onto the page, a child will use half a bottle of PVA glue!

So, you need to have a clever trick up your sleeve to protect the glue supply. Take a small Tupperware box and cut a sponge to fit inside. Then cover the sponge in PVA and let the glue get soaked up. Put the sticky sponge box onto the table. When you need to glue something down, press it onto the sponge and stick it to the page. After arts and crafts are over, put the lid back on and keep it for the next time.

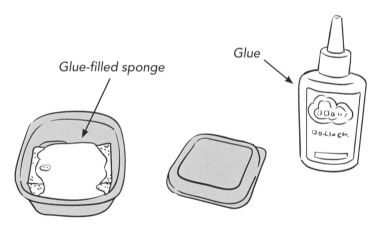

Glue-filled sponge

Glue

DIY PAINT BRUSHES

Mark making is great fun for little ones. Instead of forking out for expensive brushes you can make your own with this hack.

Attach different fabrics and materials to the end of clothes pegs by trapping them in the jaws of the peg. Use bits of sponge, netting, scouring pads, felt and whatever you have to hand. These are so simple to make that your children will enjoy making them too.

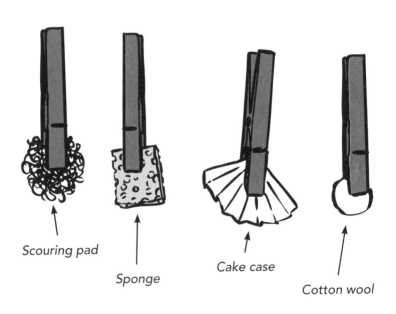

Scouring pad

Sponge

Cake case

Cotton wool

MESS-FREE PAINTING

This next hack is great for when you want to do some painting with your child but can't face clearing up afterwards.

Squirt some paint into a sealable bag (the large kind you use for freezing food) and tape the bag to a window, making sure it's well sealed. Your child can then create a masterpiece with their hands, and you can put your feet up and admire their handiwork. This hack is fun, educational, creative *and* mess-free!

Sealable freezer bag (God help you if it isn't sealed)

Modern art masterpiece

THINKING INSIDE OF THE BOX

Here's a hack that will keep your child entertained for hours. All you need is a little imagination, some crayons and a large cardboard box, preferably big enough for your child to sit or lie comfortably inside.

Ask your child to decorate the box, inside and out, however they want to. It's only a cardboard box to you, but to them it could be a castle, a spaceship or whatever their imagination allows!

Imagination running wild

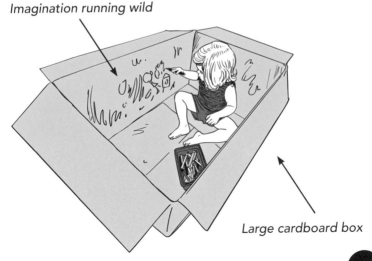

Large cardboard box

GARDEN ART

This hack is ideal for getting creative on a warm summer's day.

Stretch out an old white sheet onto the grass – pin it in place with tent pegs if there's a breeze – and let your child paint the sheet like a giant canvas. They'll be entertained for hours while getting a good dose of vitamin D.

Old sheet

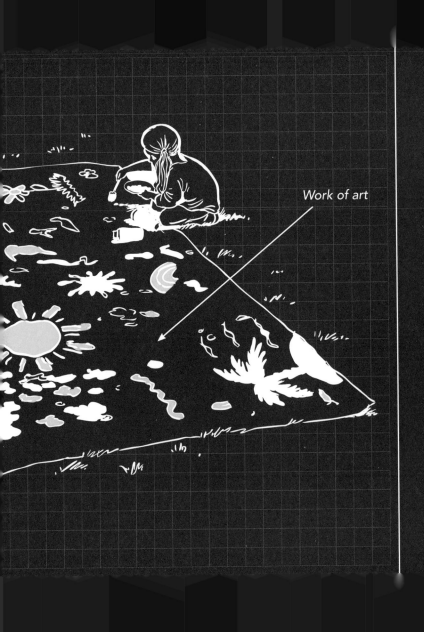

Work of art

REPURPOSED COT

This simple DIY hack is an ingenious way to get more use out of something you might otherwise throw out. When your child has outgrown their cot, repurpose it into an activity desk.

The base of the cot becomes the table-top. You could go one step further and paint the base with blackboard paint, so your child can draw on it over and over again with chalk.

Child's old cot

Child's first desk

CRAZY HOME-MADE HAT

Improve your child's hand-eye coordination with this fun hack.

Take a plastic colander and a pack of pipe cleaners and encourage your child to create a crazy hat by threading the pipe cleaners through the holes. By using their hands in this way you are preparing them for being able to hold a pen and learning to write (and possibly to be more dextrous at mischief – but let's not go there!).

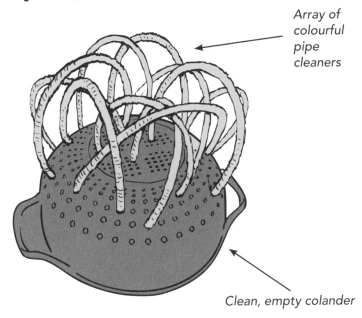

Array of colourful pipe cleaners

Clean, empty colander

ENDLESS PAPER

Children are prolific when it comes to drawing, but constantly buying paper can become expensive. Here's a cheap way to sate your child's endless need for something to draw on.

Buy a large roll of packing paper from an office supply company and suspend the roll on a wall so the child can pull the paper down, much like a toilet roll. If you're handy when it comes to DIY, create a simple frame to hold the paper in place with lengths of wood or MDF. Attach the frame to the wall, making sure there is a gap at the top and bottom to feed the paper through.

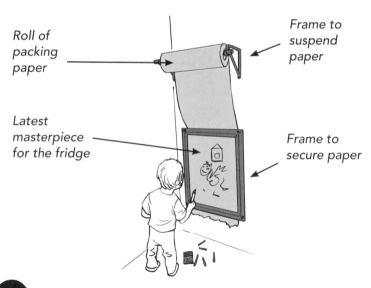

Roll of packing paper

Frame to suspend paper

Latest masterpiece for the fridge

Frame to secure paper

FUN HACKS

Time to be the fun parent and show them you still know how to have a good time with this selection of fun, sometimes educational, hacks.

MAGIC BANANA MESSAGES

Did you know bananas have the power to convey secret messages? So, presuming your child doesn't hate bananas, here's a hack to brighten their day.

Grab a toothpick or something sharp, and scratch a message onto the skin of the banana. It won't look like much at first but give it an hour or two and the peel will turn brown, revealing your message as if by magic! Perfect for kids' lunchboxes.

'Magic' banana

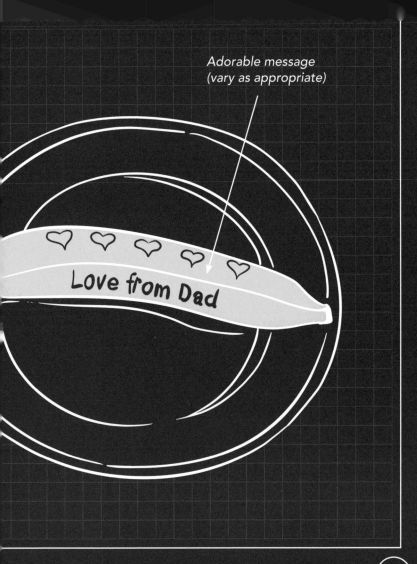

SMART SHOWER

Make your child a little smarter every time they shower with this little hack. (It doesn't involve waterproofing their textbooks or anything like that.)

Go out and buy a shower curtain with the map of the world on it so they can scrub up on their geography as they scrub themselves!

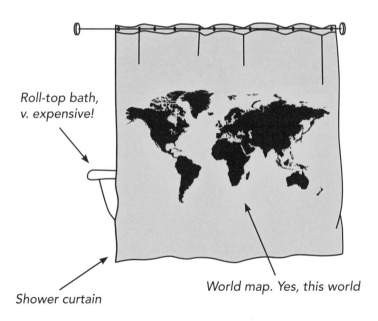

Roll-top bath, v. expensive!

Shower curtain

World map. Yes, this world

TOILET TUBE SPEAKER

Too cheap to buy speakers for your child's phone? Why would you when a toilet roll tube works just as well. Alright, that's a bit of a stretch. But it does amplify the sound pretty well.

Cut a slot big enough to fit your phone in and stop the tube from rolling away by sticking some drawing pins into the bottom to act as feet. Not only have you amplified the sound but you've also made a docking station.

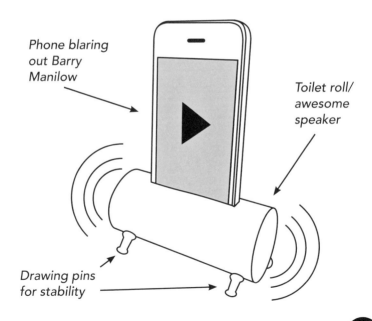

Phone blaring out Barry Manilow

Toilet roll/ awesome speaker

Drawing pins for stability

BUG FUN

If you're camping with your children and they want to stay out after the sun has gone down, they'll need some extra light – this hack shows you how to get it.

Impress the kids by putting a water bottle on top of your phone to create a makeshift lantern. The light from the screen will turn the water bottle into a light so you can see in the dark while simultaneously attracting moths! Great news if your child loves mini-beasts!

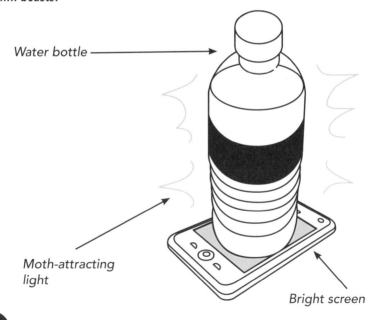

Water bottle

Moth-attracting light

Bright screen

THE FRISBEE THROW

We've all seen athletic types parading about on the beach with a Frisbee, looking cool. But they can never throw the Frisbee quite right, can they? Here's your chance to beat them at their own game.

To throw a Frisbee correctly remember to use the same action as you would when you whip a towel (come on, we've all done it!). The Frisbee will now fly as straight as an arrow. Congratulations, you are now part of their elite club.

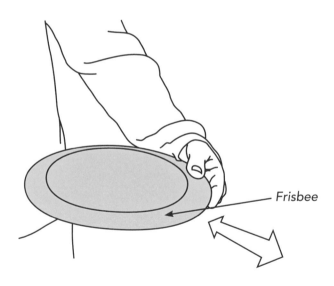

Frisbee

INSTANT TABLE HAMMOCK

Next time you want your little one to take a nap but they aren't playing ball, try this clever hack.

Tie a large bedsheet around the top of a table to create a fun mini-hammock they'll be itching to climb into. Just a few health and safety notes: test your construction for safety, and don't tie the hammock too high off the ground. A bedsheet can only hold so much weight so this is best for small kids only. (And do not, I repeat, DO NOT climb in there yourself!)

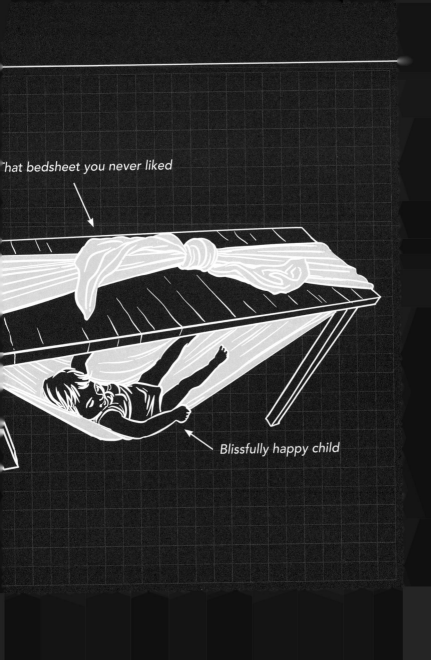

GLOW-IN-THE-DARK BOWLING

It might be inconceivable to some, but the game of tenpin bowling can be made better (in fact, this pretty much goes for anything): make it GLOW IN THE DARK!

Here's what you'll need: six glow sticks, six plastic water bottles (labels removed) full of water, a ball heavy enough to knock down your bottles (a basketball or football usually does the trick), and a pen and paper to keep score. Pour a little water out of each bottle and pop a glow stick inside each one. Set up your 'bowling pins' in a triangle formation and take it in turns to try to knock them down with your ball. Unlike going to a commercial bowling alley, this activity won't require much outlay and you get to wear your own shoes!

Water-filled bottles

Glow sticks

DIY STAIRWAY SLIDE

A houseful of bored children cooped up at home is every parent's worst nightmare. This hack will keep them entertained for hours and will turn you into the coolest parent alive.

Remember that cardboard box your rowing machine came in (yes, the one that's gathering dust in the spare bedroom)? You can use it to build an awesome stair slide. Here's how: flatten the box and tape the cardboard to the stair wall using masking tape. Pile pillows and blankets at the bottom to make a soft landing pad and then let the super-happy, fun-sliding times begin! Just remember to give the children a turn.

Shock-proofed child

Thick, snag-free cardboard slope

Arrows indicate recommended direction for use

SWING AND SWIG

We all like to be able to relax with a glass of something in the garden, but it's hard to keep the family entertained at the same time. Here's a trick that lets you put your feet up and enjoy a drink, while your little one has a whale of a time on the garden swing.

All you need is a long piece of rope. Tie the rope to the base of the swing then run the rope to your sun lounger; sit, pull the rope and swig your beer in one graceful motion.

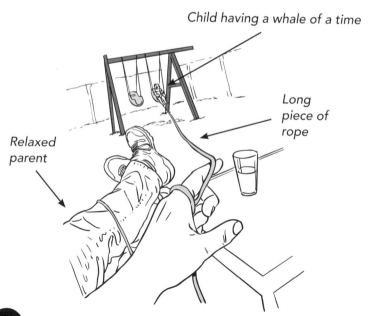

Child having a whale of a time

Long piece of rope

Relaxed parent

MONSTER VAPOURISER

If your child struggles to sleep believing there is a monster under the bed or in the dark, then this hack is for you.

Get a spray bottle and fill it with water. Decorate the outside of the bottle with glass paint; perhaps get your child to draw the monster. Then, before they go to bed, spray away the monsters so you can all enjoy a good night's sleep.

Spray bottle

Monster-
banishing
formula

SCAVENGER HUNT

Are you constantly losing things – your purse, car keys, headphones, reading glasses? You may need specialist help - but for now, try this hack.

Don't fret, the next time you lose one or several items, enlist your children to search for them for you by making it into a fun scavenger hunt. Sketch out the items on a piece of paper so they know exactly what they're looking for and give a prize for every item that they find.

List of missing items

CARD GAMES MADE EASY

Card games can be great fun but children sometimes struggle to hold all their cards without dropping them (most likely because they have sticky fingers). This simple hack will keep their cards in place.

Use an egg carton and upend it, then make slits in each of the bumps. Place the cards in the slits and your child can see all their cards in one go while keeping them out of view from other players.

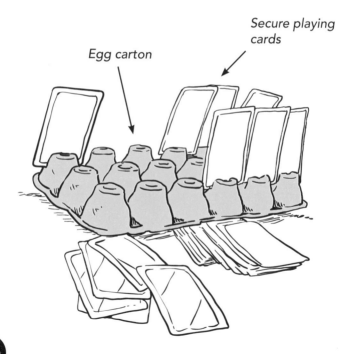

Secure playing cards

Egg carton

RANDOM HACKS

These hacks are just a little bit 'out there', but they have their merits. Besides: you've earned your stripes now as a Life Hack Hero so you deserve to know how to play for hours on the games console without your children taking over!

CHORE TRACKER

Here's a fun way to track your child's chores (what do you mean you don't give them chores?!).

Create a racetrack on a magnetic blackboard. Divide up the track into different chores, such as 'tidy your room', 'do your homework' or 'lay the table' and every time they complete a chore they move further round the track with a model car. Every time they complete a full circuit of chores, they get a new car for their collection.

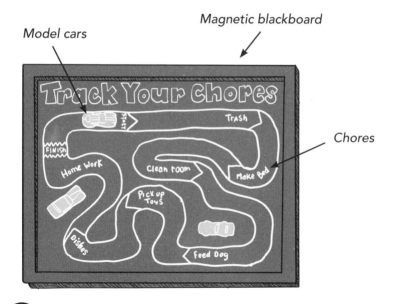

Magnetic blackboard

Model cars

Chores

I WANT, I WANT!

Shopping with children can be taxing, especially when they see so many things that they want. With this hack you won't go bankrupt and you'll have loads of ideas for what to get your child for Christmas.

The next time they see something that they simply can't live without, get your phone out and take a picture of it and say to your child that you'll send the picture to Father Christmas. Provided they still believe in Father Christmas, this should do the trick.

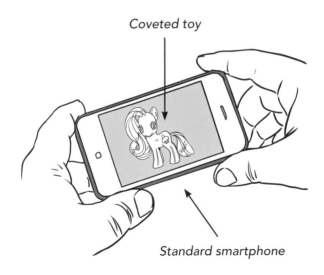

Coveted toy

Standard smartphone

FUN WALLET

Why is it that small children would much rather play with the contents of your wallet instead of the very expensive toys you've bought for them? Not only can this lead to credit cards being posted through floorboards but paper money can become ripped and unusable. The answer is to make your child their own wallet to play with.

Use a purse or wallet that you are happy to part with and fill with business cards and receipts – they'll have just as much fun playing with this wallet as they do with yours.

Wallet

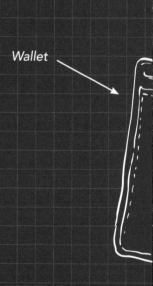

Business cards, coloured paper, definitely not credit cards

TIME CAPSULE EMAIL

This hack requires forward planning, but it's worth it!

When your children are small, create an email account for them (store the password in a *very* safe place!). Send emails to the account to mark personal milestones such as their first tooth, first steps, first day at school, etc. with photos and stories. When they're 18, give them the email address and they can relive their finest and most memorable moments.

Laptop

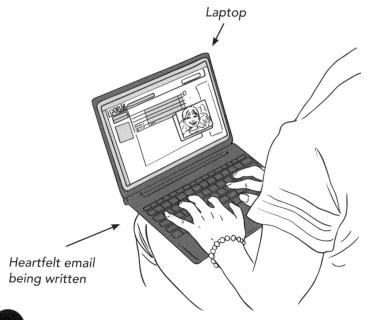

Heartfelt email
being written

CLOTHES SORTER

Do you struggle to identify which item of clothing belongs to whom when sorting the laundry? It can be tricky when your children are close in age, but here's an easy hack to make the confusion a thing of the past.

On your oldest child's clothes add one dot to the clothing labels with a permanent marker, then add two dots to your second oldest child's clothes, and so on. With a quick glance of the label on the collar or waistband you will know instantly who it belongs to.

Dots made with marker pen

Item of clothing

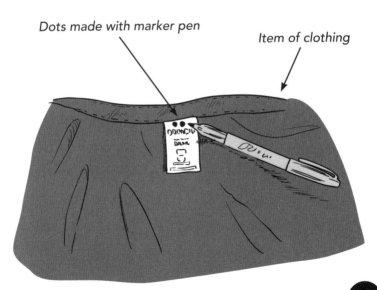

MOULD-FREE RUBBER DUCKY

Floating plastic bath toys are far from fun when they start to spew black mould, which can be hazardous for your child – here's a hack to help you avoid this potential health hazard.

Use a hot glue gun* to seal up the holes at the base of the toy so the water can't get inside them.

*Hot glue guns are cheap to buy from hobby and craft stores.

Glue gun

Faithful rubber ducky

Troublesome hole

GAMING FUN FOR ALL THE FAMILY

Children love to do what you're doing, especially when it comes to zapping aliens on the games console. This hack allows them to join in without messing up your progress.

Rather than banishing your kids from the room, or waiting until they're asleep, give them a games controller that's not plugged in and you can play together!

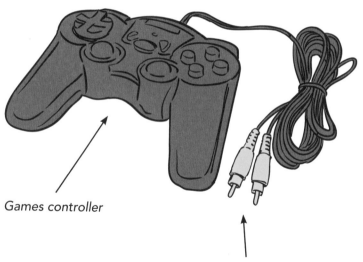

Games controller

Crucial element remaining unplugged

FINAL WORD

Congratulations – you are now a Hacks Hero. Whatever life throws at you – or whatever your children throw at you – you can handle it. Whether it's an explosive nappy or a tantrum in the toy shop, you've got it covered!

Feel free to pass on these nuggets of genius to fellow parents, and if you have some parent hacks that are not featured in this book and think they deserve to be in print, email them to auntie@summersdale.com.

HACKS INDEX